2·7·98

Materials
in
Chemical Perspective

Materials
in
Chemical Perspective

K. L. Watson BSc PhD ARIC

Lecturer, Department of Civil Engineering
Portsmouth Polytechnic, Portsmouth
England

A HALSTED PRESS BOOK

JOHN WILEY & SONS
New York

First published in 1975 by
Stanley Thornes (Publishers) Ltd.,
17 Quick Street,
LONDON N1 8HL

Published in the U.S.A. by
Halsted Press,
a Division of
John Wiley & Sons Inc. New York.

Library of Congress Cataloging in Publication Data

Watson, Keith Leslie.
 Materials in chemical perspective.

 'A Halsted Press book.'
 Includes index.
 1. Chemistry, Physical and theoretical.
 2. Materials. I. Title.

QD453.2.W38 1975   541   75-6508
ISBN 0-470-92232-X

Printed and bound in Great Britain

Contents

tion and brittle fracture. Plastic deformation. Deviation of real materials from the ideal force/distance curve.

Preface

Basic engineering materials tend to be classified as metals, ceramics and polymers. This is undoubtedly a useful division but, as it stands, it can lead the student to view each different class of materials in isolation from the others. This book sets out to examine the nature of these basic classes against a background of elementary chemistry so that, from the outset, the student can see his materials in better perspective.

The approach is based on the idea that the cohesive forces which operate in all materials are chemical in origin. A simple picture of the structure of atoms leads to an appreciation of the nature of the different forms of chemical bonding forces that, in turn, are seen as providing the cohesion which maintains the internal structure of materials and enables them to resist applied loads. The nature of a particular material depends on the type, or types of chemical bond that operate within it—this idea then provides a logical basis for the division of materials into the three classes and for a discussion of the broad differences between them in terms of their internal structure and mechanical behaviour.

The revision examples at the end of the book have been either taken or adapted from recent B.Sc. and H.N.D. materials science examination papers set in the Department of Civil Engineering at Portsmouth Polytechnic and I am most grateful to the Polytechnic authorities for allowing me to use them here.

Many colleagues, both past and present, have been of great assistance in a variety of ways and I am indebted to them; in particular, I would like to thank Dr. Tony Cox for his valuable advice. I would also like to express my thanks to the publishers for their help and encouragement, and finally to my wife, both for her patience and for preparing the final typescript.

K. L. Watson

1.

Introduction

We are all familiar with materials like glass, metals, rubber and plastics and, from experience, generally know what to expect when we handle them or accidentally drop them on the floor. For this reason we tend to accept, without question, the way that they behave.

For instance, we know that if we drop a china teacup or a glass tumbler onto a hard floor it is likely to break, but a metal spoon or a polythene bowl is unlikely to suffer much damage in the same circumstances. We say that materials like glass and china are *brittle* because they are fragile and easily broken; on the other hand we tend to regard most ordinary metals and many plastics, particularly those like polythene, as being *tough* because they are generally more difficult to break.

Unfortunately, in everyday conversation there is a tendency to confuse the terms *tough* and *brittle* with the terms *strong* and *weak*. The strength of a material is really a measure of the amount of force needed to break it and although strong materials can be tough, as in the case of steel, they are often quite brittle; for example, we know that glass generally behaves in a brittle way and that it can easily be shattered but, later on, we shall see that it is intrinsically very strong. On the other hand, polythene is quite weak in terms of the force needed to pull it apart; but it is tough in that it can put up with a lot of domestic wear and tear, and so it is useful for making things like washing-up bowls.

Another important property of materials is their *stiffness*. For instance, wood is much stiffer than rubber; wood is therefore useful for making floor-boards and bookshelves, which must be rigid, but rubber is useful for making elastic bands and tyres which must be flexible.

So we have identified three important criteria which help us to describe the behaviour of any material; these are stiffness, strength and toughness.

Over the last fifty years or so, a great deal of progress has been made in understanding the scientific reasons for these various aspects of the behaviour of different materials. This book sets out to explain, in straightforward terms, some of the basic science which helps us to understand the fundamental reasons for the familiar ways in which ordinary materials behave.

Many of us have at least heard of the idea that all matter, whether it is in solid form or liquid or even gas, is made of *atoms*. And these are extremely small; for instance a straight row of four million copper atoms placed side by side will only be just over a millimetre long. Despite their size, a great deal is known about the structure of atoms;

we shall take a brief look at this in the next two chapters and we shall see how differences in atomic structure lead to the various chemical elements (like hydrogen, carbon, oxygen, chlorine and so on) which are the basic ingredients of matter. But after this, we shall find that the central thread running through the book is the nature of the cohesive forces which bond atoms together. By understanding the nature of these forces we shall be able to view both the internal structure and the behaviour of different types of materials in proper perspective.

The idea that there are such cohesive forces in materials may be a new one to some readers; however, we can illustrate their existence by thinking about the properties of water. Water consists of hydrogen atoms and oxygen atoms; but hydrogen and oxygen are both gases and water, which is a liquid, behaves quite differently from either of them. The reason for this is that, in water, hydrogen and oxygen are chemically combined together and therefore lose their individual identities; two hydrogen atoms are linked with one oxygen atom to give a water *molecule* which we can represent by H—O—H or, in its more familiar form, H_2O. Therefore, it is not really enough to say that water consists of hydrogen atoms and oxygen atoms; it is more accurate to say that it consists of molecules, each of which contains two hydrogen atoms joined to one oxygen atom. So already we can see that there must be some cohesive force which holds the atoms together in the water molecule.

But the picture is still not complete because we know that water molecules have a tendency to stick together. This tendency is of course particularly obvious in the case of ice, where the molecules stick to one another so strongly that they form a rigid solid. However, there is also an appreciable cohesive force between the molecules in the liquid state. This is illustrated by the tendency for water to form spherical drops; even if a water drop is suspended it does not disintegrate—it behaves as though there is some invisible skin holding it together and, if it is released, it still tends to remain intact as it falls. Clearly there is some attractive force between the molecules causing them to form a coherent drop. We can of course disrupt this attractive force by raising the temperature of the water to the boiling point so that it turns to steam; but if we cool the steam it condenses back to form the coherent liquid once again.

So there would appear to be two kinds of cohesive force operating in water. Firstly there is the *primary bonding force* which holds the atoms together to form individual molecules; but there is also a *secondary bonding force* which operates between the molecules themselves.

In fact we can analyse the coherent nature of all liquids and solids in terms of internal attractive forces; and we shall examine this idea in much more detail after we have discussed the nature of atoms. But, in the meantime, we should make a point of realising that these cohesive forces are in fact electrical in nature. Because of this, we shall be making reference to forces of electrical origin throughout the book; we should therefore give them some thought at this stage to make sure that we understand something of their nature from the outset.

Many of us are probably familiar with the idea that, if we rub certain materials with a cloth, they become capable of attracting small objects like scraps of paper and hairs and, although perhaps less well-known, even of bending a thin stream of water running from a tap; we say that these materials have become electrically *charged* by being rubbed. This phenomenon has been recognised for hundreds of years, if not longer; furthermore, we know that there are two kinds of electrical charge, *positive* and *negative*, depending on the type of material and the type of cloth used.

What happens, in fact, is that in rubbing them together the material becomes charged one way and the cloth the other. This in itself is of interest to us because it suggests that, in some way, matter contains negative charge and positive charge which are normally present in equal amounts, and therefore balance one another, so that no overall charge is detected. But, by rubbing the material with the cloth, we appear to have transferred negative charge from one to the other leaving behind the equivalent surplus of positive charge (or vice versa of course). In the next chapter we shall see that there is indeed both positive and negative electrical charge in matter but, before we go on to this, we shall find it useful to briefly consider what happens when charged bodies are placed close to one another.

This is summarised by saying that like charges repel each other but opposite charges attract. So if we place two positively charged bodies close together then a repulsive force arises between them which tends to push them apart; and the same thing happens if we use two negatively charged bodies. But if we place a positively charged body close to one that is negatively charged then they are mutually attracted to one another—and, as we shall see, the consequences of this are enormously important.

But finally, before we leave this introduction and move on to think about the structure of atoms, we need to know a little about the nature of *energy*. Energy is defined as the capacity for doing work; it has a variety of forms and can be converted from one form to another. For instance, early steam engines were used to pump water out of mines; in this case, heat energy was used to perform work in raising water. But generating hydroelectric power is a similar sort of process operating in reverse; water running downhill is used to generate electrical energy which, in turn, can be used to produce heat.

There are several forms of energy that are going to be of particular interest to us.

One of these is *potential energy* which a body possesses by virtue of its position, such as its height above the ground for instance. The potential energy of a weight suspended above the ground can be used, as in driving a clock, by allowing the weight to fall. This is analogous to the hydroelectric generator where in fact the potential energy of the water is converted into electricity; we could of course just as well drive an electric generator with the falling weight.

A further example of potential energy can be shown by placing two oppositely charged bodies close together but not in contact. There is then an attractive force between the charges and they possess potential energy due, in effect, to the distance between them; work can be done by allowing them to move towards one another just as work can be done by allowing a raised weight to fall.

Kinetic energy is another form of energy that will be of interest to us. A moving object possesses kinetic energy due to its motion. For instance, a cricket ball lying on the ground cannot do work; yet if it is moving at high speed it can cause a great deal of damage—in other words it can do work, in breaking windows for example, simply because it is moving and possesses kinetic energy.

The interconversion of different forms of energy can be illustrated by thinking about the potential and kinetic energy of a toy car. Figure 1.1 shows a car standing on a track bent into the form of a valley. As long as it remains standing at the top of the hill above the valley then it has maximum potential energy; but it has no kinetic energy because it is stationary. If the car is given a gentle push over the edge, it will begin to accelerate down into the valley under its own weight; as it accelerates its kinetic energy increases but, at the same time, its potential energy decreases as it approaches the bottom—at the bottom it has no potential energy but now its kinetic energy is at a maximum. What has happened is that, in running down the hill, the car has traded one form of energy for another. But now the process begins to operate in reverse because, once it has reached the bottom of the valley, the car starts to run on

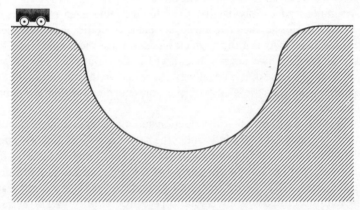

Figure 1.1: The interconversion of potential energy and kinetic energy—if the toy car is pushed over the edge it will roll down the hill under its own weight and then oscillate up and down the sides of the valley as it trades potential energy for kinetic energy and vice versa.

up the other side; and immediately it begins to slow down as it starts to trade its kinetic energy back for potential energy once again. When it has no kinetic energy left it will come to a standstill but then, of course, it will begin to roll back down into the valley once more; and this cycle of events will continue to be repeated as the car oscillates up and down the sides of the valley trading potential energy for kinetic energy and vice versa. In practice, the oscillations will soon die down and the car will come to a halt at the bottom; this is because mechanical energy is lost from the system, mostly due to friction in the wheel bearings—this lost mechanical energy is in fact converted into heat.

Another form of energy that we shall meet is *electromagnetic radiation*, and this includes light. Many of us are probably familiar with the idea that light can be dispersed with a prism to produce the well-known *spectrum* of colours ranging from red at one end to violet at the other; and it is this spectrum that we see in rainbows when sunlight is dispersed by raindrops. The dispersion of light can be thought of in terms of spreading it out into its component wavelengths—the red end of the spectrum corresponds to relatively long wavelengths and, in going through the intermediate colour range towards the violet end, the wavelength becomes progressively shorter. So a particular wavelength of light is characterised by its colour; furthermore the energy associated with electromagnetic radiation is dependent upon its wavelength—the shorter the wavelength of light, the greater the energy. In fact, the visible part of the spectrum that we can detect with our eyes is only a small part of the total spectrum of electromagnetic radiation; as the wavelength increases we pass through the infra-red region into the microwave and radio wave regions—and as the wavelength decreases from the visible we pass through the ultra-violet into the X-ray and then the gamma-ray regions.

We shall encounter energy in other forms later on in the book but, for the time being, we know enough about it to be able to go on to discuss the structure of atoms.

2.
The Structure of Atoms

Before we can discuss atomic structure we need to understand the nature of the building blocks from which atoms are made. This chapter will give a picture of the structure of atoms in terms of three types of fundamental particle—these are the *electron*, the *proton* and the *neutron*. We shall begin by briefly considering some of the broad evidence for the existence of these particles rather than follow a rigid historical account of their discovery.

The electron is the lightest of the three particles and it is negatively charged*; furthermore, it is responsible for carrying electric current through metal wires. If electrons are removed from one end of a wire and fed in at the other, then there will be a movement of electric charge along its length—and it is this movement of charge which constitutes the electric current.

To study electrons in isolation they must be persuaded to leave the metal, and one way of doing this is by heating it. For example, this principle is used in electronic valves. In its very primitive form, the electronic valve consists of a glass bulb under vacuum; a metal plate is sealed inside the bulb together with a metal filament which, like a light bulb filament, can be heated by means of an electric current. When the filament is hot it emits electrons and, if the plate is given a positive charge, the negatively charged electrons will travel from the filament to the plate—a current will therefore be carried through the vacuum by a beam of electrons.

The properties of electron beams were studied in some detail in the last century, and it was from these studies that an understanding of the nature of the electron emerged. For instance, it was shown that a small paddle wheel can be turned by a beam of electrons in just the same way that a water wheel can be turned by a stream of water; from this it could be seen that the electron possesses mass. We have already seen that electrons in the primitive valve are attracted towards a positively charged plate because of their negative charge; it was also found that, if a voltage is applied across two parallel metal plates, and an electron beam is passed between them, then the beam is deflected—its path is bent towards the positive plate and away from the negative. Furthermore, it is a basic fact of electromagnetism that a wire which carries a current will experience a force and tend to be deflected in a magnetic field; because a beam of electrons in effect constitutes an electric current, it was found to be deflected in a magnetic field too.

---

* For readers who are interested in actual values, the mass of the electron is 9.109×10^{-31} kilograms, and the charge is 1.602×10^{-19} coulombs.

Experiments of this kind involving the deflection of electron beams, together with other techniques, were used to measure the mass and charge of the electron. Electrons were produced in various ways from different materials and, irrespective of their source, the measured values of their mass and charge were found to remain the same—the electron therefore appeared to be a universal constituent of matter.

But matter is generally electrically neutral so there must be some other constituent, which is positively charged, to balance the negative charge of the electron; this is **the proton.**

Again in the last century, it was found that gases can be persuaded to conduct an electric current if large enough voltages are applied to them. This can be done at low pressure in a sealed glass tube fitted with two metal electrodes. When a large enough voltage is applied across the electrodes then the gas begins to glow (like the neon tubes used in advertising signs). Under these conditions positive particles can be detected which travel towards the negative electrode, in the opposite direction to electrons; however, these particles have their origin in the gas itself—but, like electrons, they can be studied by deflecting them with electric and magnetic fields.

The mass of these positive particles is very much greater than that of the electron and, furthermore, it varies depending on the type of gas being investigated. The lightest is detected when hydrogen (the lightest of the elements) is used, and this is the particle that we call the proton. Its mass is nearly two thousand times the mass of the electron* and its charge is exactly equal to the electron charge but opposite in sign, i.e. positive. In fact the mass of the proton is only very slightly less than the mass of the hydrogen atom from which it can be obtained; and the difference corresponds to the mass of the electron. From this we can deduce that the hydrogen atom contains one electron and one proton; by applying a high voltage across the hydrogen gas we tear the electron away from its proton—in scientific terms, we say that we have *ionised* the gas.

But, with other gases, heavier positive particles are detected; and, as with hydrogen, they are only very slightly less than the masses of the atoms from which they originate. This suggests that all atoms contain electrons and protons and, although these are normally present in equal numbers the atoms can be ionised—that is to say, electrons can be torn away from them to leave positively charged atoms, or *positive ions* as they are called.

In fact, the number of electrons (and also protons of course) characterises an atom as being that of a particular element, and this number is called the *atomic number*. The first three columns of Table 3.1, on page 16 show the elements (with their chemical symbols) which correspond to the atomic numbers from one to thirty-six. To take an example, the atomic number of chlorine is seventeen so we know that the chlorine atom contains seventeen electrons and seventeen protons. Similarly, we can see that the iron atom contains twenty-six electrons and twenty-six protons.

But now we come to a difficulty. We might expect to be able to calculate the mass of an atom from its atomic number by adding together the masses of all the electrons and protons which it contains. However, apart from the case of hydrogen, we find the

---

* The mass of the proton is 1.672×10^{-27} kilograms.

calculated mass is smaller than the actual mass measured in the laboratory. For instance, from what we have seen so far, we should expect the chlorine atom to have a mass seventeen times that of the hydrogen atom. In fact, in a normal sample of chlorine, we find a mixture of two types of atom; one is thirty-five times as heavy as the hydrogen atom and the other thirty-seven times as heavy—nevertheless, they are both chlorine atoms.

The cause of this discrepancy is the third fundamental particle that we noted at the beginning of the chapter—**the neutron.** The neutron has a slightly greater mass than the proton* but it possesses no charge. As far as we are concerned in this book, it contributes to the mass of the atom but it has no effect on its chemical behaviour. In light atoms, the number of neutrons tends to be about the same as the number of protons; the relative number of neutrons increases as the size of the atom increases and heavy atoms generally contain about one and a half times as many neutrons as protons.

In the chlorine-35 atom there are seventeen electrons and seventeen protons; therefore there must be eighteen neutrons to make up the difference in mass—and in the chlorine-37 atom there must be twenty neutrons. (N.B. Atoms of the same element which have different masses in this way are called *isotopes*.)

To summarise, we have seen that the electron is the negatively charged component of the atom but contributes very little mass to it. Most of the mass is provided by the neutrons and protons. The proton is the positively charged component of the atom and, because atoms are electrically neutral, the number of protons contained in them must be equal to the number of electrons.

But so far we have given no thought to how electrons, protons and neutrons are arranged within the structure of the atom. The clue to this came towards the beginning of this century as a result of experiments in which thin metal foils were bombarded with *alpha-particles*. These particles, which are emitted by some radioactive isotopes, consist of two protons united with two neutrons so they possess an appreciable mass and have a double positive charge. The experiments showed that most of the alpha-particles passed more or less straight through the metal foils but that a very small number were strongly deflected.

One conclusion drawn from this was that, since most of the alpha-particles passed through the foils, then matter consists very largely of empty space; but the fact that some were strongly deflected pointed to the existence of *nuclei* in matter which contain most of its mass and which are positively charged so that positive alpha-particles would rebound from them because of repulsive forces. The rarity of these events indicated that the nuclei are widely separated and are therefore very small indeed.

In fact, as this suggests, the nucleus of the atom contains all its neutrons and protons and therefore provides a central core possessing all the positive charge and virtually all the mass. But where are the electrons?

Early models of the atom viewed it as an analogy with the solar system; the electrons were regarded as revolving around the nucleus in orbits just as the planets revolve in their orbits around the sun. Each electron, being negatively charged, is at-

---

* The mass of the neutron is 1.675×10^{-27} kilograms.

tracted towards the positively charged nucleus but this attractive force is opposed by an outward centrifugal force resulting from the electron's circular path—the electron therefore settles into a stable orbit at a distance from the nucleus where the inward attractive force and the outward centrifugal force exactly balance one another. The total energy of the electron in its orbit consists of its potential energy (due to its distance from the nucleus) and its kinetic energy (due to its motion); the larger the total energy of the electron, then the larger is the diameter of its orbit.

It has been estimated that the effective diameter of atoms, i.e. taking into account the space occupied by the moving planetary electrons, is of the order of thousands of times the diameter of the nucleus. This provides us with a picture of an atom which mostly consists of empty space and through which alpha-particles can readily pass unless they happen to be on a path which leads them very close to the nucleus—the electrons are unlikely to have much effect on the path of the alpha-particles because their relative mass is so small.

However, there is a fundamental difficulty. The classical laws of electromagnetic radiation suggest that an electron travelling in its circular path should be regarded as a tiny oscillator, continuously emitting electromagnetic energy. It is beyond the scope of this book to show why this is so—nevertheless the arguments are analogous to those which explain how radio waves are emitted as a result of oscillations in a transmitter aerial. The important point to note is that the electromagnetic energy which the electron should emit must be supplied by the electron itself. The consequence of this is that its energy would be reduced and so it would move in closer towards the nucleus; it would therefore follow a spiral path into the nucleus as its energy was consumed in emitting electromagnetic radiation.

The model therefore needs some modification because, if electrons really behaved in this way, then matter would simply collapse in a blaze of electromagnetic radiation.

We already know that atoms can be persuaded to emit light because, as we noted earlier, gases will glow if we apply high voltages to them—but we have to supply the energy first. Furthermore, in experiments of this kind, if we disperse the light with a prism the resulting spectrum is seen to consist of individual well-separated lines rather than the continuous rainbow spectrum seen when ordinary white light is dispersed. In the case of the hydrogen spectrum, four lines fall within the visible band of electromagnetic radiation and others can be found by using special detection techniques. But we cannot explain this emission of radiation at discrete wavelengths in terms of our simple model of the atom as it stands.

These difficulties were overcome by modifying the model in such a way that the electron can only adopt specific discrete orbits. The underlying theory supporting this step came as part of what is now seen as a revolution in theoretical physics which occurred at the beginning of this century. The impact of this on our simple picture of the atom is not very difficult to understand in principle, but it is perhaps helpful to begin thinking about it in terms of an analogy.

Comparing the earlier model of the atom with the modified one is rather like comparing a smooth slope with a staircase. If we place a ball at the top of the slope it will roll downhill trading its potential energy for kinetic energy rather as, in the earlier model, the electron would have run downhill into the nucleus trading its energy for emitted electromagnetic radiation. In the case of the staircase, the ball can only possess specific fixed levels of potential energy corresponding to each step. Similarly, in the modified model of the atom, the electron can adopt only specific fixed orbits, as in Figure 2.1, and the larger the orbit the higher the energy level of the electron.

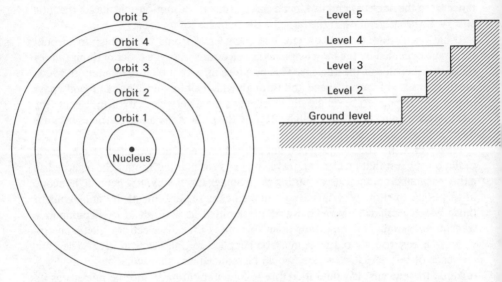

Figure 2.1: The staircase analogy—the electron can adopt only certain specific orbits in the atom which correspond to fixed energy levels, rather as the steps of a staircase correspond to fixed levels of potential energy.

The revolutionary step came in suggesting that the electron can remain in a fixed orbit without emitting electromagnetic energy. This model allows the electron to move from orbit to orbit but, depending on whether the transition is from a low energy orbit to a higher one or vice versa, then energy either has to be supplied to the atom or will be emitted by it. In terms of the analogy, the ball will possess fixed potential energy which depends on which step it is resting—but its potential energy can be increased or decreased by fixed amounts by either raising it or lowering it to other steps.

We can now see the reason for the discrete lines in the hydrogen spectrum. In supplying energy to the gas (by applying a high voltage across it) the electrons in the atoms are promoted to high energy orbits. When an electron reverts to a lower energy orbit, electromagnetic radiation is emitted which corresponds to the difference in energy between the higher orbit and the lower one. Because there are only a specific number of permitted energy levels there are only a limited number of energy transitions which can occur—the wavelength of the emitted radiation is related to the change in the energy of the electron and so only wavelengths corresponding to those particular energy transitions will be seen.

This modified model therefore explains the stability of atoms and why they do not collapse—furthermore, it leads to an understanding of the discrete lines in the spectrum.

But if the spectrum is examined very closely we find that it has a "fine structure"; what at first appeared to be individual lines are made up of a number of fine lines very close together—so there are more transitions possible than the model suggests. This led to the view that the main orbits, or "shells" as they are often called, should be divided into subshells with energy levels slightly different from one another; in pictorial terms, these subshells were originally viewed as allowing for the existence of elliptical orbits of varying form. As Figure 2.2 shows, the first shell is not subdivided but the second is divided into two, the third into three and the fourth into four. These subshells are identified by the letters *s, p, d* and *f*. This notation dates back to the early days of experimental work on spectra, but it has been retained and is still used as the basis for identifying the address of a particular electron in an atom. For instance, if the electron is in the *p* subshell of the second shell it is described as a 2*p* electron; if it is in the *f* subshell of the fourth shell it is called a 4*f* electron, and so on.

The situation became more complicated when it was discovered that the lines of the spectrum were split still further if the atoms were under the influence of a strong magnetic field. There are therefore further subdivisions of the subshells themselves and these only become apparent in the presence of a magnetic field. As Figure 2.2 indicates, the *s* subshells do not split but the *p* subshells split into three, the *d* into five and the *f* into seven. This splitting was attributed to the possible orientations that the orbits could adopt in relation to the magnetic field.

Finally it was found that even lines of this kind could be further resolved into very closely spaced pairs. This is represented by the fourth column of Figure 2.2 and was attributed to the idea that the electron can spin about its own axis, like a spinning top, in either a clockwise or an anticlockwise direction.

So we can now regard Figure 2.2 as summarising all the possible addresses that are available for electrons in the first four shells of the atom. This broad picture still stands today although the notion of electrons revolving in circular or elliptical orbits has been greatly modified. The story of this development is a complex and highly mathematical one and is beyond the scope of this book; nevertheless we shall find it helpful to briefly consider its effect on our simple picture of the atom.

The modern view of the atom dispenses with the idea of the electron being confined to fixed orbits. Instead we think of three-dimensional *orbitals*, as they are now called, which correspond to regions within the atom in which the probability of finding the electron is high. Figure 2.3 indicates the overall shape of the envelopes which enclose the regions of high probability for the electron is the 1*s*, 2*s* and 2*p* orbitals of the hydrogen atom.

The probability envelopes for the 1*s* and 2*s* orbitals are spherical in shape and are centred on the nucleus. Figure 2.2 showed us that *p* subshells subdivide into three and we can see, from Figure 2.3, that the probability envelopes for the three 2*p* orbitals are dumbell shaped; these are each centred on the nucleus and are orientated, mutually perpendicular to one another, along the *x, y* and *z* axes in the figure. This is as far as we shall need to go, but it is possible to describe the higher orbitals too. We must remember that, in each orbital, there are two addresses corresponding to the two spin states; in other words each orbital can accommodate two electrons of opposite spin.

To summarise all this, we now have a picture of the atom as containing specific ad-

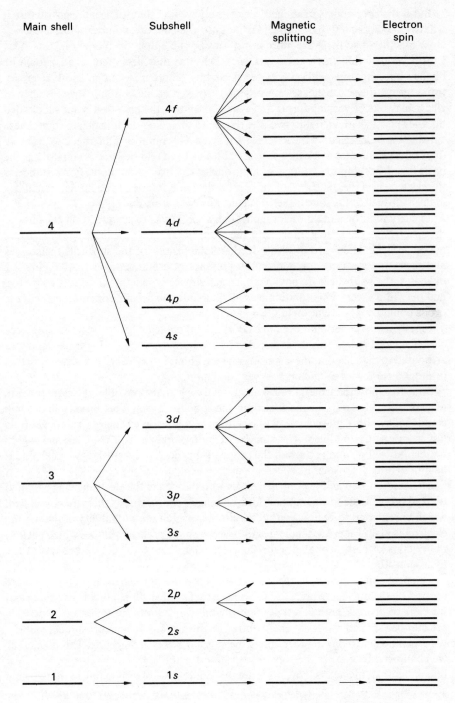

Figure 2.2: The possible addresses available for electrons in the first four shells of the atom.

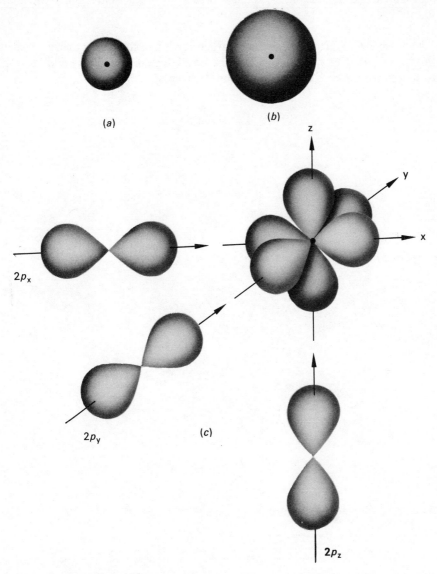

Figure 2.3: The probability envelopes for the electron in the hydrogen atom: (*a*) the 1*s* orbital; (*b*) the 2*s* orbital; (*c*) the three 2*p* orbitals—$2p_x$, $2p_y$ and $2p_z$.

dresses that can be occupied by electrons. Keeping Figure 2.2 in mind, we can see that in the first shell there are two addresses corresponding to the two spin states in the 1*s* orbital. In the second, there are two in the 2*s* orbital and two in each of the three 2*p* orbitals, making eight in all. Similarly, in the third shell there are eight in the 3*s* and 3*p* orbitals, but a further ten in the five 3*d* orbitals making a total of eighteen. There are an additional fourteen in the seven $4f$ orbitals making a total of thirty-two in the fourth shell.

Now we can go on to think about the structure of more complex atoms; we shall do this by imagining that, starting from hydrogen, we could successively add electrons to occupy these addresses that we have just been considering.

3.
The Elements and the Periodic Table

In the last chapter we saw that there are a number of addresses available for electrons in each main shell of the atom; there are two in the first shell, eight in the second, eighteen in the third and thirty-two in the fourth—and, although we did not go beyond the fourth, in fact there are more in the higher shells too.

In this chapter we shall think about complex atoms containing many electrons. We shall only consider the electrons themselves and how they occupy their addresses; but of course we should bear in mind that, in any particular atom, there must be an equal number of protons in the nucleus*.

We shall begin by considering the elements as a progressive series that starts with hydrogen and in which each successive member contains one more electron than its predecessors. We have already noted that Table 3.1 shows the first thirty-six members; furthermore, we have seen that each element is characterised by its atomic number, i.e. the number of electrons that the atom contains. But now we can see that the table also indicates the order in which the addresses are filled as we proceed from one element to the next.

In thinking about the order of filling, the important point to remember is that the addresses which are filled first are those of lowest energy. As Figure 2.1 suggested, the first main shell contains the lowest energy addresses; this shell is therefore filled first. So, in the hydrogen atom, the single electron normally occupies the 1s orbital†. But this still leaves a vacancy in the 1s orbital for a second electron having opposite spin to the first; Table 3.1 shows that when this vacancy is filled we have the ground state electronic structure of the helium atom—and the first main shell is complete.

In the case of lithium, the second main shell begins to fill; so now we have to decide whether the third electron enters the 2s subshell or one of the three 2p subshells. We can do this by referring to Table 3.2 which shows the subshells in order of increasing energy; each circle in the table represents an orbital which is capable of containing two electrons of opposite spin. As we can see, the 2s subshell corresponds to an energy level which is lower than the 2p subshell; so, in the ground state of lithium, the third electron will occupy the 2s orbital as indicated in Table 3.1. The fourth electron,

---

*   * In this book we shall not be concerned with neutrons, nevertheless we should remember that these are also present in the nucleus.
    † This represents the *ground* state of the hydrogen atom. We must not forget that the electron can be promoted to a higher level, i.e. an *excited* state, if we supply the necessary energy to it.

ELEMENT	SYMBOL	ATOMIC NUMBER	1s	2s	2p	3s	3p	3d	4s	4p
Hydrogen	H	1	1							
Helium	He	2	2							
Lithium	Li	3	2	1						
Beryllium	Be	4	2	2						
Boron	B	5	2	2	1					
Carbon	C	6	2	2	2					
Nitrogen	N	7	2	2	3					
Oxygen	O	8	2	2	4					
Fluorine	F	9	2	2	5					
Neon	Ne	10	2	2	6					
Sodium	Na	11	2	2	6	1				
Magnesium	Mg	12	2	2	6	2				
Aluminium	Al	13	2	2	6	2	1			
Silicon	Si	14	2	2	6	2	2			
Phosphorus	P	15	2	2	6	2	3			
Sulphur	S	16	2	2	6	2	4			
Chlorine	Cl	17	2	2	6	2	5			
Argon	Ar	18	2	2	6	2	6			
Potassium	K	19	2	2	6	2	6		1	
Calcium	Ca	20	2	2	6	2	6		2	
Scandium	Sc	21	2	2	6	2	6	1	2	
Titanium	Ti	22	2	2	6	2	6	2	2	
Vanadium	V	23	2	2	6	2	6	3	2	
Chromium	Cr	24	2	2	6	2	6	5	1	
Manganese	Mn	25	2	2	6	2	6	5	2	
Iron	Fe	26	2	2	6	2	6	6	2	
Cobalt	Co	27	2	2	6	2	6	7	2	
Nickel	Ni	28	2	2	6	2	6	8	2	
Copper	Cu	29	2	2	6	2	6	10	1	
Zinc	Zn	30	2	2	6	2	6	10	2	
Gallium	Ga	31	2	2	6	2	6	10	2	1
Germanium	Ge	32	2	2	6	2	6	10	2	2
Arsenic	As	33	2	2	6	2	6	10	2	3
Selenium	Se	34	2	2	6	2	6	10	2	4
Bromine	Br	35	2	2	6	2	6	10	2	5
Krypton	Kr	36	2	2	6	2	6	10	2	6

Table 3.1: The ground state electronic configurations of the first thirty six elements.

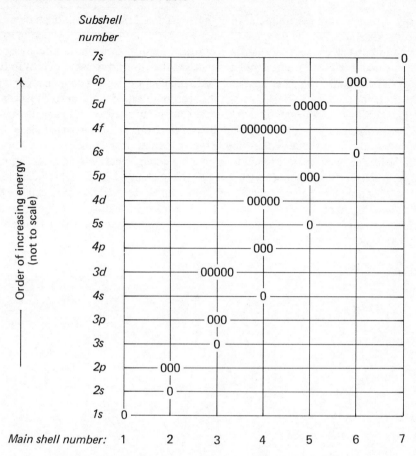

Table 3.2: Atomic subshells shown in order of increasing energy.

in beryllium, will complete this orbital and then the $2p$ subshell will begin to fill, starting with boron. We can follow the filling of the $2p$ subshell in Table 3.1 and we can see that neon marks the completion of the three $2p$ orbitals—the second main shell is then full.

But complications begin to arise when we come to the third main shell. Table 3.2 indicates that the subshell energy levels of this shell overlap with those of the fourth; in this particular case the energy level corresponding to the $4s$ subshell is below that of the $3d$. So, to begin with, the $3s$ and $3p$ subshells fill and Table 3.1 tells us that this leads to the elements from sodium to argon; but when the $3p$ subshell is complete, the $4s$ must be filled before the $3d$ can begin—and this gives us potassium and calcium. Then, starting with scandium, the $3d$ subshell begins to fill but it is not until we reach gallium that the $4p$ can begin.

This is as far as we shall need to go to illustrate how the electronic structure of the elements can be viewed in terms of filling empty subshells in order of increasing energy. An important general point to note from Table 3.2 is that, although overlapping first occurs when the 4s subshell begins to fill, after this point it is continuous.

Two important consequences arise from this. Firstly, as we have already seen, elements occur which have incomplete inner subshells—and we shall return to this point later. Secondly, the outermost main shell of any atom cannot contain more than eight electrons; and, as Table 3.2 shows, the reason for this is that *d* and *f* subshells of any main shell will not begin to fill unless there are electrons present in a higher main shell. As we shall now see, the actual number of electrons in the outermost main shell is of very great importance in determining the properties of an atom.

In the last chapter we noted that, by supplying enough energy to an atom, we can tear an electron away from it to leave it positively charged. In fact it is possible to measure the amount of energy needed to remove an electron and we shall call this the *ionisation energy* of the atom*. Figure 3.1 shows the ionisation energy of the first twenty elements arranged in order of increasing atomic number. We can use this graph to draw some interesting conclusions.

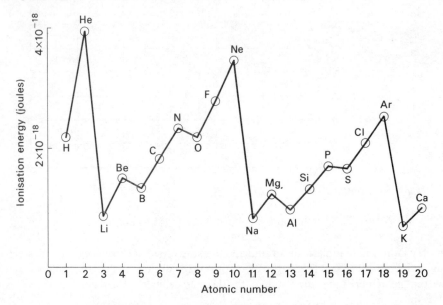

Figure 3.1 : Ionisation energies of the first twenty elements.

Firstly we notice that helium (chemical symbol He), neon (Ne) and argon (Ar) show high ionisation energies. The atoms of these elements are therefore resistant to the removal of an electron, and this suggests that their electronic structures are particularly stable. On the other hand lithium (Li), sodium (Na) and potassium (K) are particularly unstable since it is relatively easy to remove an electron from these. In building up from lithium to neon and from sodium to argon the ionisation energy increases but it drops suddenly from helium to lithium, from neon to sodium and from argon to potassium.

---

* Strictly speaking we should call it the *first ionisation energy* to imply that we are only removing one electron.

Cross-reference to Table 3.1 shows that, in the particularly stable electronic structures of neon and argon, the outermost main shell contains eight electrons, i.e. the *s* and *p* subshells are full; and we already know that this is the maximum number of electrons that an atom can contain in its outermost shell. These elements which have this complete octet of outer electrons are called the *inert* gases (or sometimes *noble gases*) since they are all gases which are extremely reluctant to combine chemically with other elements. Apart from these, there are heavier inert gases called krypton (atomic number 36), xenon (54) and radon (86). Helium is also an inert gas but it differs from the others in only possessing two electrons in its outermost shell; as we already know, these are in the first shell, which is therefore full, and so there is no possibility of it containing a further six *p* electrons.

If we think about the elements which have one more electron than the inert gas structure (i.e. lithium, sodium and potassium) then Table 3.1 reminds us that the additional electron begins a new shell. Figure 3.1 suggests that, in this configuration, the outermost electron is particularly loosely held in the atom because so little energy is needed to remove it; as we shall see later, this makes the atom readily amenable to chemical combination with other atoms. These elements, which only have one electron in the outermost shell, are called the alkali metals and the later members of the series not shown in Figure 3.1 and rubidium (atomic number 37), caesium (55) and francium (87).

We have already noticed that there is an increase in ionisation energy if we go along the series from an alkali metal to the next inert gas. For instance, in building up from lithium to neon, the ionisation energy increases by a factor of four; this corresponds to an increase in atomic number from three to ten—so the outermost electron in lithium is attracted towards the nucleus by the positive charge due to three protons whereas the outermost electrons in neon are attracted by ten. Clearly an electron which is held by ten positive charges will be more tightly bound to the atom than an electron only held by three; therefore more energy is needed to remove an electron from a neon atom than from lithium.

Figure 3.1 also shows that, within the stable inert gas family, there is a decrease in ionisation energy as the atoms become heavier; the ionisation energy for helium is higher than for neon which, in turn, is higher than for argon. Although the trend is much less obvious, the ionisation energy also decreases along the series lithium, sodium and potassium—and we can see the same effect for the elements which have two electrons in the outermost shell, i.e. beryllium (Be), magnesium (Mg) and calcium (Ca). In fact similar trends can be observed generally as atoms become larger.

This effect is due to the increasing screen of inner electrons shielding the outer electrons from the attraction of the nucleus; the presence of inner electrons makes it easier to tear an outer electron away from the atom against the attractive force. Taking the inert gases as an illustration, we can regard the two electrons in helium as being in direct sight of the nucleus. In the case of neon, we can think of the outermost electrons as being screened from the nucleus by the two electrons in the first shell; the outer electrons are therefore less strongly held and the ionisation energy is correspondingly less. In argon the outermost electrons are screened by the ten electrons in the inner two shells; an outer electron is therefore still easier to remove.

And, of course, the abrupt decrease in ionisation energy between an inert gas and the succeeding alkali metal is due to the sudden large increase in the number of screening electrons relative to the increase of only one proton in the nucleus.

Taking an overall view of Figure 3.1, we can see that there is *periodicity* in the electronic stability of these elements; in other words we can see that the peaks in ionisation energy occur at intervals of eight elements—and these intervals correspond to the successive filling of the outer shells with eight electrons.

In fact it is rather more convenient to set out the elements in the form of a table in which each row corresponds to one interval. This has been done in Table 3.3 for the first thirty-six elements.

The alkali metals, with one outer electron, are all contained in the first vertical column together with hydrogen—and these are sometimes called the Group I elements. The Group II elements, in the second column, all have two outer electrons. The Group III elements have three, and so on. So the number at the head of each column corresponds to the number of electrons in the outermost shell for that group of elements; the only exception is the inert gas group in column eight which, by convention, is given the group number 0.

The first horizontal row has six gaps between hydrogen and helium because of the absence of *p* electrons in the first shell; but the elements in the second, third and fourth rows correspond to the successive addition of eight electrons to the second, third and fourth shells respectively.

This table, which reflects the periodicity of the elements, is called the *periodic table*. One important point to remember is that, by rearranging the elements in this way, the ionisation energy tends to show a general increase as we go across the table from left to right; and of course this is due to the effect of the increasing positive charge on the nucleus as the number of outer electrons increases from one to eight. On the other hand, the increasing screening effect of the larger number of inner shells of electrons, as we move down the table, is reflected by the general tendency for the ionisation energy to decrease in this direction.

Before leaving the periodic table we should briefly think about the complications that are introduced by the overlapping of subshell energy levels between the main shells. Up until argon (atomic number 18), Table 3.3 is perfectly straightforward. The complications begin in the fourth row.

We have already seen that, in potassium and calcium, the inner $3d$ subshell remains empty while the $4s$ subshell fills. Furthermore, after calcium, the $4p$ subshell does not begin until after the $3d$ is completed. So, although the ten elements from scandium to zinc really belong in between calcium and gallium, they do not readily fit into the fourth row as it stands; instead they form an inner series, called the *transition elements*, which fit in between Groups II and III as shown in Table 3.3.

Table 3.2 shows that there is continuous overlapping after this and so there are still further series of transition elements. We therefore need to construct a rather more complex version of the periodic table to include all the elements, and this is shown in Table 3.4; we can see that more complications begin with the lanthanide series which corresponds to the filling of the $4f$ subshell before the $5d$ can be completed. However, we are not concerned with the detailed aspects of the full table here but simply with

Group No:	I	II	III	IV	V	VI	VII	O
	₁H (1)							₂He (2)
	₃Li (2.1)	₄Be (2.2)	₅B (2.3)	₆C (2.4)	₇N (2.5)	₈O (2.6)	₉F (2.7)	₁₀Ne (2.8)
	₁₁Na (2.8.1)	₁₂Mg (2.8.2)	₁₃Al (2.8.3)	₁₄Si (2.8.4)	₁₅P (2.8.5)	₁₆S (2.8.6)	₁₇Cl (2.8.7)	₁₈Ar (2.8.8)
	₁₉K (2.8.8.1)	₂₀Ca (2.8.8.2)	₃₁Ga (2.8.18.3)	₃₂Ge (2.8.18.4)	₃₃As (2.8.18.5)	₃₄Se (2.8.18.6)	₃₅Br (2.8.18.7)	₃₆Kr (2.8.18.8)

TRANSITION ELEMENTS

₂₁Sc (2.8.9.2)	₂₂Ti (2.8.10.2)	₂₃V (2.8.11.2)	₂₄Cr (2.8.13.1)	₂₅Mn (2.8.13.2)	₂₆Fe (2.8.14.2)	₂₇Co (2.8.15.2)	₂₈Ni (2.8.16.2)	₂₉Cu (2.8.18.1)	₃₀Zn (2.8.18.2)

Table 3.3: The arrangement of the first thirty six elements in the periodic table.

(*N.B.* The atomic number for each element is shown beside its chemical symbol and the number of electrons in each main shell is shown in brackets below.)

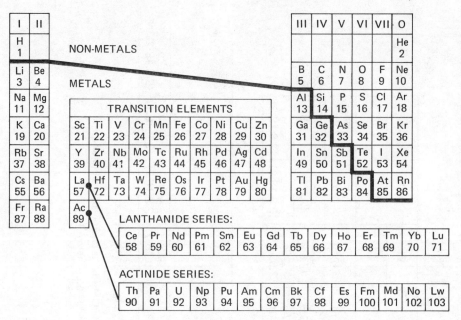

Table 3.4: The periodic table of the elements.

(*N.B.* The atomic number of each element is shown below its chemical symbol.)

the idea that its structure can still be viewed in terms of the general order of filling indicated in Table 3.2.

This is as far as we need to go to see how the overall structure of the periodic table arises. But before we leave this part of the discussion we should note that the elements below and to the left of the heavy line in Table 3.4 are metals and those above and to the right of it are non-metals; in fact some elements close to the line can show behaviour that is partly metallic and partly non-metallic and the exact position of the division in some parts of the table can be argued.

This brief discussion of the periodic table has illustrated how the elements can be arranged in a logical way depending on their electronic structures. But the periodic table, in its original form, was formulated on the basis of the observed behaviour of the elements; and this was long before the electronic structure of atoms was understood and before some elements had even been discovered.

The idea that the behaviour of an atom is fundamentally related to its electronic structure is enormously important. Now that we have some understanding of the electronic structure of atoms, we can go on to see how they interact with one another; and this will help us to understand the nature of the cohesive forces which make them stick together to form useful materials.

4.

The Combination of Atoms—Chemical Bonding

In the last chapter we saw that it is relatively difficult to remove an electron from an inert gas atom; a complete outer octet of electrons, or a pair in the case of helium, represents a particularly stable configuration. It is therefore perhaps not surprising that atoms which are not inert gases tend to achieve these stable electronic configurations if they can. Sometimes they can do this by reacting with other atoms in such a way that they either gain or lose electrons.

For instance Table 3.1 shows us that if a sodium atom could get rid of its outermost electron, for example by giving it to some other atom, then its electronic configuration would become the same as for neon. Of course, in doing this, the sodium atom does not become a neon atom; its nucleus, and hence its number of protons, remains unaltered—so the loss of one electron means that it becomes a positive sodium ion. We can represent this process in chemical shorthand as follows $Na \rightarrow Na^+ + e^-$. Na represents the neutral sodium atom, Na^+ the positive sodium ion and e^- the electron.

To take another example, chlorine is a Group VII element and therefore has seven electrons in its outermost shell. Table 3.1 shows that, if a chlorine atom is able to acquire an electron from some other atom, it can then achieve the stable electronic configuration of the inert gas argon. This means that the chlorine atom then has one extra electron and is therefore negatively charged; in other words it has become a negative ion which, in this particular case, is called the chloride ion. Again we can represent the process in chemical shorthand, this time using the chemical symbol Cl to represent chlorine: $Cl + e^- \rightarrow Cl^-$.

In fact both these tendencies are satisfied in sodium chloride, or common salt as it is more generally known: in effect, we can regard the single outer electron of the sodium atom as having been transferred to the outer shell of the chlorine atom. Both atoms have therefore become ions having inert gas configurations. But this has had a profound effect; after all, chlorine is a highly poisonous gas and sodium is a soft metal that is so chemically active that it even reacts violently when it comes into contact with water—on the other hand, salt is used for seasoning food and exists in the form of hard brittle crystals. Why should the location of a single electron be so important?

We will begin by considering why salt should be a crystalline solid. The observation that it exists as a solid means that there must be some cohesive force which holds the crystal together. In fact, this force is simply a result of the attraction between sodium ions and chloride ions due to their opposite charges; and this type of chemical bond, which causes ions to stick together to form crystal structures, is called the *ionic bond*.

Because the nature of the ionic bond is an important factor in determining the properties of many materials, we need to think about it in some detail. The first point to remember is that we can regard ions, with their inert gas configurations, as being rather like tiny charged spheres. If two oppositely charged ions are a long way from each other then there will only be a relatively small attractive force between them; but as they are brought together then this force will progressively increase. (To take an analogy, the force between two magnets is relatively small when they are a long way apart but it increases as they are brought closer together.) The upper dotted line in Figure 4.1 illustrates how the attractive force between two oppositely charged ions

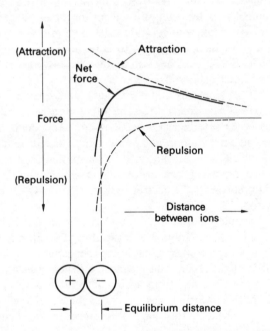

Figure 4.1: The balance of forces in the ionic bond. The net force curve is obtained by combining the attraction and repulsion curves.

varies with distance. If the two ions are placed close together, the attractive force between them will tend to make them move towards one another until they come into contact. But, before we go any further, we should really decide what we mean by the word *contact* since ions are nebulous objects which mostly consist of empty space.

The lower dotted line in Figure 4.1 shows that, when the ions come very close together, a strong repulsive force develops which opposes the attractive force; but, as the figure suggests, this repulsion only becomes really important when the distance between the ions is relatively small. In fact, it occurs when their outer electron shells approach very closely and begin to repel one another; so, in simple terms, we can regard the outer shell as behaving rather like a spherical elastic skin around each ion which prevents it from being squeezed into the other.

Figure 4.1 shows that there is a balance between the forces of attraction and repulsion in the ionic bond. The net force at any particular distance between the ions can be found by combining the component attractive and repulsive forces at that point—and the overall picture is given by the net force curve shown in the figure. We can see that the attractive force predominates when the ions are relatively far apart; but, although it increases as they move nearer to one another, the repulsive force begins to increase still more rapidly as their outer electron shells come close together—and then the repulsion effect becomes the predominant component of the net force.

The figure shows that there is a point at which the attractive and repulsive forces exactly balance each other so that the net force is zero; and this point represents the *equilibrium distance* which the ions will adopt spontaneously if they are allowed to come to rest by themselves. The net force curve shows that, if we try to push the ions closer together than their equilibrium position, the balance between the attractive and repulsive forces is disturbed with the result that a net repulsive force arises; this then provides a resistance to our efforts to push the ions closer together. Similarly, the curve shows that any attempt to pull the ions apart from their equilibrium position will be resisted by the net attractive force which develops between them when we upset the balance in this direction.

So now we can begin to understand how ionic crystals are held together and why they are rigid solids, i.e. why they can resist our efforts to squash them or to pull them apart. Figure 4.2 provides us with a simple two-dimensional picture that gives us some idea of how the ionic bonds operate collectively in a crystal structure. We can see that, in this model, each positive ion is surrounded by a cage of four negative ions, and vice versa; so each ion is held firmly in its place by four ionic bonds formed with its oppositely charged neighbours. Clearly, ions of like charge will tend to keep away

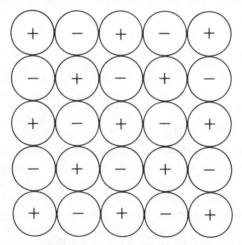

Figure 4.2: A two-dimensional representation of an ionic crystal structure.

from each other because of their mutual repulsion; and, in effect, we can regard ionic crystal lattices in terms of packing spheres together in such a way that oppositely charged spheres come into contact with one another but spheres of like charge remain apart—and the result is an extended and regular crystal structure. Of course, to represent real crystal structures, we must think in terms of the three-dimensional packing of ions, but we shall elaborate on this in more detail in Chapter 6.

Up until now we have only thought about ions which possess a single charge. For Group II elements to achieve an inert gas configuration they must lose two electrons as, for instance, in the case of magnesium: $Mg \rightarrow Mg^{2+} + 2e^-$. Conversely, taking oxygen as an example, Group VI elements must gain two electrons: $O + 2e^- \rightarrow O^{2-}$. So magnesium oxide is an ionic material which has a crystal lattice in which O^{2-} ions are packed around Mg^{2+} ions, and vice versa of course. Complications arise if the magnitude of the charge on the ions is different; for example in calcium fluoride, the calcium ion (Ca^{2-}) has a double positive charge whereas the fluoride ion (F^-) only has a single negative charge—there must therefore be twice as many fluoride ions as calcium ions in the crystal structure.

So we have now seen that the ionic bond involves the complete transfer of electrons from one atom to another. But it is possible for atoms to achieve stable configurations by sharing electrons. This results in another type of chemical bond, called the *covalent bond*, and a simple example of it is given by the hydrogen molecule. So far we have thought of hydrogen in terms of single atoms which each contain one electron so that, in the ground state, the $1s$ orbital is half-filled. But in fact hydrogen normally prefers to exist as H_2 molecules or, more explicitly, as H—H molecules in which two atoms are joined together by a covalent bond. In this case we can regard each atom as having contributed its single electron to form a pair which is shared between them, in effect giving them both the full complement of two electrons that is associated with the helium configuration.

Figure 4.3 shows that we can view the bond as resulting from the overlapping of the two half-filled orbitals to give a single joint orbital, containing both electrons,

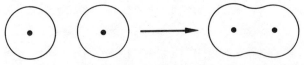

Figure 4.3: The formation of the hydrogen molecule can be viewed in terms of the overlapping of the half-filled orbitals of two individual atoms to give a joint orbital that contains both electrons.

which embraces the two atoms and bonds them together. As before, it is possible to analyse the bond in terms of attractive and repulsive forces. Firstly, the electrons will spend much of their time in between the two nuclei; negative electrons situated between the two positive nuclei will obviously exert an attractive pull on both of them and will therefore, in effect, bond them together. On the other hand, if the nuclei are pulled in too close together a strong repulsive force will arise between them. There will therefore be an equilibrium distance which corresponds to the point at which the attractive force is exactly balanced by the repulsive force.

This gives us a picture of the hydrogen molecule as a stable entity which is able to maintain a separate existence on its own. Under normal conditions hydrogen molecules tend to remain independent of one another; in other words they exist as a gas rather than coalesce to form a cohesive liquid or solid.

Now we can begin to see why chlorine normally exists as a gas. We already know that the chlorine atom, like hydrogen, is one electron short of its neighbouring inert gas configuration; so again there is scope for two atoms to share two electrons, one from each, to form a complete orbital joining them together. Chlorine therefore normally exists as separate molecules, each containing two covalently bonded atoms; the molecules tend to remain independent of one another and therefore normally exist in the form of a gas.

An interesting point to note is that a chlorine atom can form either ionic or covalent bonds; and, as we have seen, the choice depends on the nature of the other atom with which it forms the bond. For instance, a sodium atom can readily lose an electron so that both it and the chlorine atom can achieve inert gas configurations, and the resulting bond is ionic. But a chlorine atom cannot achieve an inert gas configuration by losing an outer electron because it will still have six left; so two chlorine atoms do not form an ionic bond together but, instead, they share a pair of electrons to form a covalent bond.

Carbon is another element that forms covalent bonds; furthermore, it has particular interest for us because it is the basis of most rubbers and plastics. For this reason we shall think about it in some detail. Table 3.4 shows that carbon is a Group IV element. So, to form an ion having an inert gas configuration, the carbon atom would need to either lose or gain four electrons; but this is very difficult to do and carbon tends to form covalent molecules by sharing four pairs of electrons with other atoms.

One way that it can do this is to combine with four hydrogen atoms to form methane, which normally exists as a gas. Figure 4.4a shows that, in the methane molecule, each hydrogen atom is linked to the carbon atom by a pear-shaped orbital

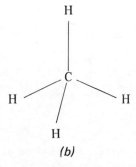

(a)   (b)

Figure 4.4: The methane molecule: (a) the four orbitals, joining the hydrogen atoms to the central carbon, spread apart as far as possible and (b) the resulting molecule is shaped like a triangular-based pyramid with a hydrogen atom at each corner and the carbon atom in the centre.

and that the molecule has a definite shape; the four orbitals repel one another and so they tend to spread as far apart from each other as possible. As the figure indicates, the result of this is that the molecule has an overall shape like a triangular-based pyramid with a hydrogen atom at each corner and the carbon atom at the centre; and we describe this arrangement as *tetrahedral* because the molecule is shaped like a tetrahedron with the orbitals pointing into the corners. Figure 4.4*b* shows the molecule in a more schematic form that is convenient for our purposes; each atom is represented by its chemical symbol and each covalent bond by a line.

At this point it is worth briefly interrupting the discussion to emphasise some important distinctions between the nature of ionic and covalent bonds that are now becoming apparent. We have already seen that the ionic bond does not lead to discrete molecules as such; instead we think of ionic crystals in terms of charged spheres which pack together under the influence of the attractive force due to opposite charges. So the ionic bond is not specific to two particular ions; it operates between any two that are oppositely charged and in contact with one another—furthermore it is not directional and we can pack as many negative ions as we like around each positive ion, and vice versa, so long as no ions of like charge touch each other. On the other hand, atoms can only form a specific number of covalent bonds which depends on the number of electrons in their outermost shell; hydrogen and chlorine atoms can form one, carbon can form four and oxygen (with six outer electrons) can form two. And of course this limitation to the covalent bonding capacity of an atom leads to the formation of discrete molecules capable of independent existence. Another distinguishing feature of the covalent bond is that it exists specifically between the two atoms which share a pair of electrons in a joint orbital; and because orbitals tend to point in specific directions, covalent molecules have definite shapes. Having emphasised these distinctions we can now go on to look at some more carbon-based molecules in better perspective.

Figure 4.5*a* shows the structure of the ethane molecule, which contains two carbon atoms; again these both form four covalent bonds, one with each other and three

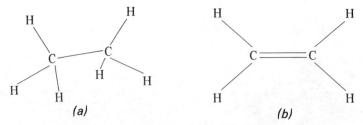

Figure 4.5: (*a*) the ethane molecule, and (*b*) the ethylene molecule.

with hydrogen atoms. And, as we might expect, ethane normally exists as a gas. An interesting property of this molecule is that one end can rotate relative to the other, rather like a propeller (see Figure 10.2*a*, page 96); and in Chapter 10 we shall see that rotation about carbon–carbon bonds in this way has an important bearing on the properties of rubber.

Ethylene is another gas that consists of molecules containing two carbon atoms; but Figure 4.5*b* shows that these are joined together by a *double bond* in this case. For our purposes we can simply regard the double bond as resulting from the sharing of two pairs of electrons between the atoms so that they are bonded together by two joint orbitals—and, of course, this means that each carbon atom can only bond with two hydrogen atoms, rather than three as in ethane. But it also means that the double bond is shorter than the single bond; the combined effect of the two joint orbitals is to pull the carbon atoms closer together—and it also restricts bond rotation about their common axis so that the molecule tends to remain locked into position in such a way that all its component atoms lie in the same plane, i.e. the plane of the paper in the figure. Furthermore, the double bond is stiffer than the single bond; a larger force is needed to stretch the carbon atoms apart—and more energy is needed to completely separate them.

But so far we have only thought about small molecules. Carbon can form larger covalent structures too; methane (CH_4) and ethane (CH_3CH_3) are the first two members of a series of chain-like molecules which are based on a backbone of carbon atoms held together by covalent bonding—the next two members are propane ($CH_3CH_2CH_3$) and butane ($CH_3CH_2CH_2CH_3$). We shall meet this series again when we discuss rubbers and plastics in Chapter 10 but, in the meantime, we can see the chain-like form of this type of molecule in Figure 10.1 on page 95.

But covalent bonding can lead to extended three-dimensional crystal structures too. One example is diamond, which is a form of pure carbon. We shall think about the diamond structure in more detail later on but, for the time being, Figure 6.15*a* (on page 57) shows us how each carbon atom is held in its fixed position in the crystal lattice by covalent bonds formed with its four neighbouring carbon atoms—these are arranged around it in a tetrahedral pattern like the hydrogen atoms in the methane molecule. Of course the figure only represents a tiny portion of the crystal structure, which extends in three dimensions.

Although we have made a rigid distinction between ionic and covalent bonds, we should note that this does tend to give an oversimplified view. It is not necessary to complicate the picture unduly, but we should be aware that ionic bonds can be regarded as having varying degrees of covalent character and covalent bonds as having varying degrees of ionic character. We will start by briefly thinking about covalent character in ionic bonds.

Ions are not all the same size. For instance, the oxide and fluoride ions (i.e. O^{2-} and F^-) are considerably larger than the sodium, magnesium and aluminium ions (i.e. Na^+, Mg^{2+} and Al^{3+}); and, in fact, there is a general decrease in size as we go along the series in that order. All these ions have the electronic configuration of neon but the number of protons in their nuclei varies; the oxide and fluoride ions contain eight and nine protons respectively, and the sodium, magnesium and aluminium ions contain eleven, twelve and thirteen. So the decrease in the size of the ions corresponds to an increase in the positive charge on the nucleus; there is a tendency for negative ions to be comparatively large and we can regard the outer skin of electrons as being relatively loosely held by the nucleus because there are fewer protons—on the other hand, the outer electrons in positive ions tend to be relatively tightly held by the larger number of protons, and the electron skin tends to be more compact. Furthermore, there is an overall tendency for ions to become larger as we proceed down the periodic table because the greater number of inner electrons leads to increased screening of the outer electrons from the nucleus.

So how does all this lead to covalent character in the ionic bond? Let us imagine that we have two oppositely charged ions in contact; one is a small positive ion which we shall regard as having a compact hard electron skin, and the other is a large negative ion with a relatively loose and floppy skin. When the ions are close together the loose electron skin of the negative ion can be distorted by the attraction due to the positive ion and, in effect, be displaced towards it; the outer electrons in the negative ion will therefore tend to spend more time on the side closest to the positive ion, i.e. in between the two nuclei. In other words, there is a tendency towards sharing of the electrons between the two atoms and so, in effect, the bond possesses some covalent character. The smaller and more highly charged the positive ion the greater is its ability to distort negative ions in this way; and the larger and more highly charged the negative ion, then the more capable it is of being distorted.

But how do covalent bonds possess ionic character? We can illustrate this by thinking about the water molecule. The oxygen atom contains six electrons in its outermost shell and so it can form two covalent bonds with hydrogen atoms to give the H—O—H molecule. Each hydrogen atom is bonded to the oxygen atom by a joint orbital containing two electrons which spend most of their time in between the two nuclei. However, the oxygen nucleus contains eight protons whereas the hydrogen nucleus contains only one; the result of this is that the negatively charged electrons will tend to be more closely associated with the oxygen nucleus than the hydrogen—the oxygen end of each hydrogen–oxygen bond will therefore tend to assume a negative bias leaving the hydrogen end with a positive bias. And this tendency towards the separation of positive and negative charge, where unlike atoms are bonded together, can be regarded as giving the covalent bond some ionic character.

So now we have some understanding of both the ionic and the covalent bond. And, taking sodium chloride and diamond as examples, we have seen that both types lead to crystalline solids; furthermore, covalent bonding can also lead to the formation of small molecules, such as hydrogen and water, and to larger ones too. But there is still a third primary bond that we have not yet considered; this is called the *metallic bond,* and it is responsible for the cohesive force which holds metal crystals together. We shall take sodium as an example to give us a simple picture of how this type of bond works.

As we already know, the sodium atom can fairly readily lose its outer electron to form a positive ion. In its simple form, the metallic bond in sodium can be viewed as the result of a collection of sodium atoms giving up their outer electrons to become positive ions—the electrons then become free and form a *sea* of negative charge which surrounds the ions, as illustrated in Figure 4.6. In fact the electrons wander around at random in between the ions and so the figure really represents an instantaneous snapshot which shows the position of the electrons at a particular moment in time. There will be an attractive force between each electron and its surrounding ions and so it will tend to bond them together. But, since the electrons are moving around, they will continually be changing their neighbouring ions; so, in effect, we have an overall picture of positive sodium ions which form a crystal lattice that is held together by a glue of negative electrons that are in constant random motion.

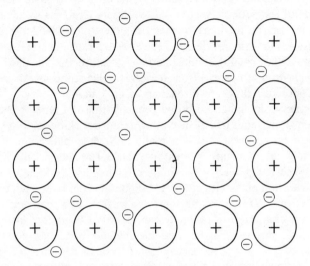

Figure 4.6: The metallic bond. The positive metal ions can be regarded as being held together by the attractive forces due to the free electrons in between them at any instant.

We can see that our model of the metallic bond differs fundamentally from the ionic and covalent bonds in that the bonding electrons are not localised; they operate collectively and are not confined to specific nuclei. Furthermore, we can regard the metal ions as being indirectly bonded to each other so, as we shall see later, they tend to form crystal structures in which they pack together as closely as possible with no directional constraints. Again, repulsion effects begin to become important when the ions come very close together and, as with the ionic and covalent bonds, we can think of the metallic bond in terms of a balance between attractive and repulsive forces.

One important feature of the metallic bond is that the metal ions do not all have to be of the same kind—we can have a mixture of different ions all held together in the same crystal structure. This explains how we can form alloys in which different metals, such as copper and nickel for example, are combined together.

Furthermore, the freedom of movement of the bonding electrons in metals results in high electrical conductivity; in Chapter 2 we noted that electrons are responsible for carrying electric current through metals—and now we can see that it is their freedom to move that enables them to do this so easily.

The free electrons in a metal are not confined to specific energy levels but exist in continuous bands, or ranges of energy. An enormously large number of energy transitions is possible and so most metal surfaces can both absorb and emit virtually all the wavelengths of light; and this gives metals their characteristic lustre.

So, to very broadly summarise our discussion so far, we have seen that there are three types of primary bonding force which cause atoms to stick together. Elements on the left hand side of the periodic table contain more electrons than their closest inert gas structure; so, on their own, these atoms tend to lose their outer electrons to form positive ions which are bonded together in a crystalline structure by a negative sea of electrons—and we call these elements metals. Non-metallic elements on the right hand side of the periodic table need to acquire electrons to achieve the configuration of their nearest inert gas; on their own, these atoms tend to share electrons between themselves to form covalent molecules which range in size from two atoms, like chlorine, up to the virtually infinite diamond structure. But metallic atoms can, in effect, transfer their outermost electrons to non-metallic atoms and the resulting ions form a crystal structure that is held together by ionic bonding.

So this summary highlights the broad principle that the type of chemical bond formed between atoms tends to depend on their relative positions in the periodic table. But we should bear in mind that it is really an oversimplification. For instance, hydrogen is shown as a Group I element in Table 3.4 because it only has one outer electron, but it is certainly not a metal; on the other hand, it is one electron short of an inert gas configuration and so it can form a single covalent bond and, to this extent, behaves like a Group VII element. Furthermore we have seen that ionic and covalent bonds often have characteristics that we can regard as being intermediate between the two, and later on we shall see that the metallic bond can possess covalent characteristics; but, in fact, a more detailed analysis of the intermediate nature of chemical bonds shows that even this can often be related to the relative positions of the bonded atoms in the periodic table.

We have now examined the nature of the cohesive forces which are involved in metals, in ionic crystals like sodium chloride, and in covalent crystals like diamond. But small covalent molecules form solid materials too, although usually only at relatively low temperatures; a familiar example is water, which freezes at 0°C. However, we have seen that the bonding capacity of the component atoms in covalent molecules is used in constructing the molecules themselves; therefore there are no spare electrons available for forming further primary bonds. Why, then, should these molecules stick together to form solid coherent materials?

The answer is that there are secondary bonding forces which do not rely on atoms gaining, losing or even sharing electrons; instead they depend upon the mutual attraction between molecules, and between atoms, which occurs as a result of *polarisation,* i.e. the separation of positive and negative charge within the molecules or atoms involved.

In discussing covalent bonds with ionic character, we saw how permanent polarisation occurs when two unlike atoms are bonded together. Attractive forces will arise between permanently polarised molecules if they are arranged in such a way that the positive part of one comes close to the negative part of another. A particularly important form of this type of bond is the hydrogen bond, which occurs in water.

Let us start with ice. Figure 4.7a shows the shape of the water molecule, and we can see that it is not unlike the methane molecule in its general form. Firstly there are two orbitals (unshaded in the figure) which are each involved in bonding a hydrogen atom to the central oxygen atom; and, as we saw earlier in this chapter, the outer end of each of these orbitals has a positive bias because the oxygen nucleus tends to draw the electrons away from the hydrogen nucleus. But the oxygen atom still has four outer electrons left over which are not involved in forming bonds. These occupy two *non-bonding* orbitals (with two electrons in each) which are shaded in the figure; since they only encompass the oxygen nucleus they are shorter than the bonding orbitals, which are stretched out because of the hydrogen nucleus at their outer ends—therefore, in effect, the non-bonding orbitals represent relatively dense regions of negative charge in the molecule.

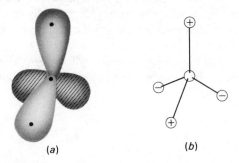

Figure 4.7: The water molecule: (*a*) showing the two bonding orbitals (unshaded) and the two non-bonding orbitals (shaded) and (*b*) showing the permanent polarisation of the molecule in schematic form.

So the water molecule has two orbitals that are positively biased and two that are negatively biased—and we can represent it schematically in the form shown in Figure 4.7*b*. In a collection of water molecules we would expect each individual to orientate itself with respect to its neighbours; each of its own orbitals will tend to point towards oppositely charged orbitals of surrounding molecules. The result of this is that attractive forces will tend to hold the molecules together, as illustrated in Figure 6.16 on page 58. The collective effect will be an overall cohesive force which tends to hold the entire mass of molecules together. In the next chapter we shall consider the conditions which determine whether the molecules will form a liquid or a solid; for the time being, it is sufficient to say that the cohesive force in water and ice is basically due to the attraction between oppositely charged parts of neighbouring molecules. And this cohesive force, involving the attraction between a non-bonding orbital and an orbital bonding a hydrogen atom, is called the hydrogen bond; it is not confined to molecules where these orbitals are attached to oxygen atoms but occurs in other cases too—nevertheless, water is the most significant example as far as we are concerned in this book.

But our picture of the cohesive forces in materials is still not complete. For instance, we have no explanation of why molecules such as H—H and Cl—Cl form liquids and solids if we cool them down to low enough temperatures; clearly there will be no permanent polarisation of the covalent bond because the bonded atoms are alike and so the pair of electrons will be equally shared between them. Nor can we explain how the inert gases form liquids and solids if we cool them down far enough at sufficiently high pressure; inert gases have a complete outer octet of electrons (or pair in the case of helium) and therefore exist as single atoms—so primary bonding cannot be responsible for their cohesion, and again, nor can permanent polarisation of the kind that we saw in the water molecule. Why should these single atoms stick together?

We can derive the answer to this by looking at Figure 4.8 which represents the helium atom at different moments in time; if the electrons happen to be exactly opposite one another along a line through the nucleus (Figure 4.8a) then there is no net polarisation of the atom. But the electrons are in perpetual motion and, in a situation like the one shown in Figure 4.8b, they are not diametrically opposite one another and the atom will therefore be temporarily polarised; the other three examples (Figure 4.8c, d and e) illustrate that, in any situation other than where the two electrons are diametrically opposed, temporary polarisation will exist—furthermore, owing to the perpetual movement of the electrons this polarisation will continuously change in magnitude and direction.

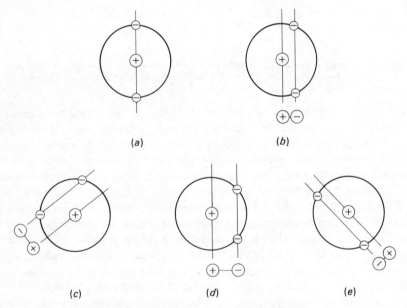

Figure 4.8: Fluctuating polarisation in the helium atom.

If there is a second helium atom nearby, then the polarisation in this will tend to fluctuate in sympathy with the first; for instance Figure 4.9 shows that if, at any moment in time, the electrons in atom A happen to be one the side closest to atom B then they will tend to repel the electrons in atom B—if they happen to be on the opposite side, the nucleus of atom A will be exposed to atom B and will therefore tend to attract the electrons accordingly. So the polarisation of the second atom will tend to be orientated similarly to that of the first, and an attractive force will therefore arise between the two. The polarisation in the second atom will, in turn, tend to induce sympathetic fluctuations in the polarisation of any neighbours that it may have (including a reciprocal effect on the first atom), and so on; the result is a network of attractive forces which tend to hold all the atoms together. Although the picture is more complicated in the case of the small molecules containing two similar atoms, the origin of the cohesive force is basically the same; the electrons in adjacent molecules will tend to move in sympathy, as in the two helium atoms—and the result is an attractive force which tends to hold the molecules together.

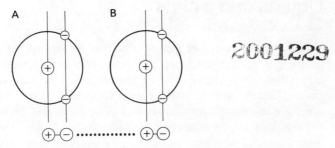

Figure 4.9: The van der Waals bond. The polarisation in neighbouring atoms tends to fluctuate in sympathy; for example, at the instant represented in the figure, the electrons in atom A are tending to repel the electrons in atom B—the temporary polarisation in each of the two atoms is then similarly orientated and so there will be a force of attraction between them.

Forces of this kind are called *van der Waals forces* (after a scientist of that name) and this type of mechanism will operate between all atoms that are close together and molecules as well. However, van der Waals forces are very weak and may not be apparent unless all other cohesive forces are absent, as in the case of the inert gases for example.

So we can see how individual molecules, and even atoms, can be held together by secondary bonding forces which do not involve the transfer or sharing of electrons but, instead, rely on polarisation—either permanent, as in the hydrogen bond, or temporary in the case of the van der Waals bond. As before, we can represent these secondary bonds in terms of a balance of forces; the attractive force which they produce is balanced by a repulsive force which arises when the outer orbitals of the molecules or atoms involved come close together. As we might perhaps expect, hydrogen bonds are stronger than van der Waals bonds but they are still significantly weaker than the primary bonding forces that we discussed earlier.

Although this chapter has given us a broad picture of the basic types of chemical bonding forces which are responsible for the existence of coherent materials, it has left us with two important questions. Firstly, it has not explained the existence of gases; in other words it has not told us why atoms and molecules do not all coalesce under the influence of these bonding forces at ordinary temperatures—furthermore, it has not told us why there should be two distinct states of coherency, i.e. liquids and solids.

5.
The States of Matter—Gases,
Liquids and Solids

At the end of the last chapter we saw that the existence of gases poses a problem. On the grounds that even inert gas atoms tend to stick together, we might perhaps expect that all matter should take a coherent form. But the observation that there are gases indicates that there must be some factor tending to oppose the cohesive forces which are responsible for the existence of liquids and solids.

There are some clues that help us to identify this factor. If we pick out the elements from Table 3.1 which normally exist as gases we find that they are either small covalent molecules (H_2, N_2, O_2, F_2 and Cl_2)* or inert gases (He, Ne, Ar and Kr); although these elements can form liquids and solids under the influence of van der Waals forces at low temperatures and sufficiently high pressure, the weak attractive forces provided by this type of bonding do not give sufficient cohesion at normal room temperatures. Secondly, the hydrogen bond is stronger than the van der Waals bond; and H_2O molecules normally exist in the liquid state. But substances in which the strong cohesive forces due to metallic, ionic and covalent bonding occur extensively are normally solid materials. All this suggests that the physical state of a particular substance at room temperature is in some way related to the strength of the cohesive forces which operate within it.

To look at this another way, the state in which a particular substance exists will depend on its temperature; for instance, H_2O molecules either form ice, steam or water depending on whether we cool them, or heat them or leave them at normal room temperature. So heat appears to be the factor which opposes the cohesive forces in materials.

To understand the differences between gases, liquids and solids let us imagine that we begin with a collection of unspecified molecules in the gaseous state. We shall then see what happens if we cool them down so that they condense to form a liquid. Finally we shall think about cooling the liquid to form a solid material.

At atmospheric pressure and 0°C, a litre of gas (just under two pints) contains nearly 3×10^{22} molecules; under these conditions the average distance between two

---

* The two atoms in the hydrogen, fluorine and chlorine molecules are bonded together by a single covalent bond. Oxygen is a Group VI element and so the two atoms in the oxygen molecule are joined by a double bond, O=O. But nitrogen is a Group V element and this molecule is held together by a triple bond, N≡N, in which the two atoms share three pairs of electrons.

neighbouring molecules is about 30×10^{-10} metre which, for fairly small ones, is very roughly ten times their size. So although the air that we breathe and walk about in seems rather unsubstantial, it is really surprisingly dense.

In the gaseous state, the molecules are in complete disorder and are free to move around in a random fashion. Under these conditions they continually bump into each other. To simplify the discussion we shall regard them as behaving rather like elastic spheres. Because of their random motion and frequent collisions the molecules will be travelling with a wide range of velocities; but their average velocity remains constant at a fixed temperature. In fact they move quite quickly, typically in the order of hundreds of metres per second at 0°C. However, their average velocity varies with temperature; if a gas is heated then, on average, the molecules move faster. We can regard the addition of heat energy to the molecules as leading to an increase in their kinetic energy.

But why do the molecules stay in the gaseous state? What prevents them from sticking together under the influence of the cohesive forces that we know to exist between them? The short answer, in simple terms, is that this is due to the effect of their relatively high velocity. If two molecules move slowly towards one another so that they collide gently then they will remain in contact because of the cohesive force between them. But if the molecules are moving very quickly, so that they collide violently, then the cohesive force will not be strong enough to restrain them from rebounding apart again.

So if we cool a gas, and reduce the thermal motion of the molecules, we shall find a temperature (which depends on the nature of the gas) at which the cohesive forces become effective in drawing them all into a cohesive state in which they are held in contact with one another—in other words, they form a liquid.

Figure 5.1 shows a two-dimensional representation of the gaseous and liquid states. As the figure suggests, gases can easily be compressed because there is so much free space between the molecules; liquids, on the other hand, are relatively in-

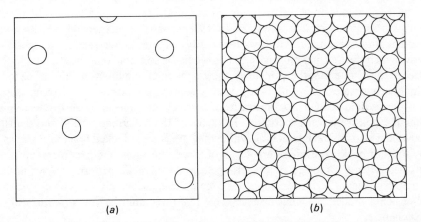

(a)   (b)

Figure 5.1: Two-dimensional models of the gaseous and liquid states: (a) in the gaseous state the molecules are free to move independently of one another, and (b) in the liquid state they are held together in a coherent mass although they are still capable of moving relative to each other.

compressible because most molecules are in contact with their neighbours. Gases will expand to fill the entire volume of their container but liquids have a determinate volume because the molecules are held together in a coherent mass.

In the liquid state, the molecules are still in continuous motion relative to one another—and for this reason liquids can flow—but the molecular movement is much more restricted. A molecule in a gas can generally travel for some distance before it collides with another; in a liquid it does little more than shuffle around and it is perpetually being jostled by its neighbours. So Figure 5.1 really represents an instantaneous picture and we should remember that, in reality, the molecules are in constant relative motion.

A molecule in the body of the liquid is uniformly surrounded by neighbours and, on average, is subjected to attractive forces equally in all directions. But a molecule at the surface experiences a net inward attractive force, towards the body of the liquid, because it only has neighbours on that side. And this tendency for surface molecules to be pulled inwards means that a liquid will tend to adjust its shape so that the number of molecules at the surface, hence its surface area, is at a minimum. The geometrical shape which possesses the least surface area for any particular volume is the sphere; so liquids tend to form spherical drops if they can, although in practice these are generally distorted by other factors such as gravity.

This tendency for liquids to minimise their surface area leads to the idea of *surface tension*. To stretch a liquid film against its surface tension, just as we do in blowing a soap bubble, work must be done—and the energy that we use to do this is converted into *surface energy* in the stretched film.

So, to summarise our discussion, we have tended to regard the liquid state as a collapsed gas; in cooling a gas until it liquefies, the thermal motion of the molecules is reduced to the point at which the cohesive force between them is able to pull them into a coherent mass—but the molecules are still in a sufficient state of thermal agitation to be able to move relative to each other.

The thermal motion of the molecules in our model of the liquid state is still dependent upon temperature; on cooling, they shuffle around less energetically and the liquid becomes more viscous, i.e. it flows less easily. Eventually we reach a temperature below which the thermal motion of the molecules is so feeble that they can no longer squeeze past one another; instead they are each confined within a cage of neighbouring molecules so that they occupy nearly fixed positions—in other words, they form a rigid solid. The cohesive force between the molecules has, in effect, overcome the relatively weak disruptive influence of their thermal motion and tends to draw them into an arrangement in which they are closely packed together. And as in any form of packing, even laying a wall of bricks, the most compact formations are obtained when the molecules are arranged in a systematic pattern, like the crystal structure in Figure 5.2 for example. Of course the figure merely shows a simple two-dimensional model; we shall discuss some real three-dimensional structures in the next chapter.

Each of the molecules still undergoes thermal motion, even in an ordered crystal structure of this kind; but this is now restricted to vibration about a fixed point in the crystal lattice. These thermal vibrations do, in fact, tend to force the molecules slightly apart; if we lower the temperature of the crystal the vibrations become less vigorous and the average distance between two adjacent molecules decreases a little—and this explains the thermal contraction that can be observed when a crystalline material is cooled and, conversely, the expansion when it is heated.

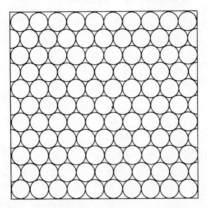

Figure 5.2: Two-dimensional model of a crystal structure; thermal motion is restricted to vibration about fixed points in the crystal lattice.

Of course it is possible to reverse the sequence of events which occurs when a gas is cooled to form a liquid and finally a solid. And to consolidate our picture of the states of matter it is perhaps worth briefly running through the process in the opposite direction.

If we begin with a crystal at low temperature, the thermal motion of its component molecules is confined to vibration about fixed points in the crystal lattice. As the temperature is raised these vibrations become more vigorous and the crystal expands. When we reach the melting point the thermal vibration is so great that the molecules become free to shuffle around relative to one another, although the cohesive forces are still able to maintain them in a coherent mass. As the temperature is raised still further, these shuffling movements become more and more vigorous. At the boiling point the thermal motion becomes sufficiently energetic to completely overcome the internal cohesion of the liquid and the molecules then fly apart quite freely and independently of one another.

This discussion of the effect of temperature on matter has necessarily been brief; it is an enormous subject and there are many aspects that we have had to ignore. We have confined the discussion to molecules relying on secondary bonding forces for their cohesion; similar arguments apply to materials which rely on primary bonding forces although these are stronger and the temperatures needed to disrupt them are generally much higher. Furthermore, we have talked about molecules as though they are simple elastic spheres and we have neglected the fact that the chemical bonds within the molecules themselves will undergo thermal vibrations too. We have thought about the transition from the liquid state to the solid without any reference to the way in which crystals actually grow. And we should also note that there are many materials which cannot be changed into a gas, or some even melted by heating because they actually decompose or burn before this can happen.

Nevertheless we can now see that there is a balance between the cohesive forces which tend to make atoms, ions or molecules stick together and the disruptive effect of thermal motion which tends to break them apart.

6.

Crystal Structures

We have seen that when the thermal energy of atoms or ions, or indeed molecules, is sufficiently low then they will pack together in ordered crystalline structures under the influence of the chemical bonding forces which operate between them. However, they must be sufficiently free to move to be able to settle down into their stable, low energy positions in the crystal lattice. If their movement is restricted whilst they are trying to crystallise into an ordered arrangement then a glassy non-crystalline structure can result; we shall discuss this idea in more detail when we think about the structure of glass in Chapter 9 but, in this chapter, we shall look at some of the geometrical arrangements that are to be found in ordered crystalline structures.

The precise nature of a crystal structure depends on the type of chemical bonding forces holding it together; so we shall find it convenient to discuss the various arrangements in terms of the type of bond that is involved. It is perhaps easiest to begin with metals.

We know that a solid metal can be regarded as a collection of positive ions bound together in fixed relative positions in a sea of electrons. And we also know that the separation between the centres of two neighbouring metal ions corresponds to the equilibrium distance on the net force/distance curve.

In discussing the crystal structure of metals, it is more convenient to replace this model of ions stuck together with electrostatic forces by a model made from hard spheres stuck together with glue. In fact, large scale models of metal crystal structures are often made in just this way using table-tennis balls; the distance between the centres of two table-tennis balls in contact then represents the equilibrium distance. As we shall see in Chapter 8, it is possible to go quite a long way in relating the properties of metals to their structures in terms of this hard sphere model; so we shall use it now to look at the three most important metal crystal structures in terms of ways in which hard spheres can be packed together to form three-dimensional arrangements.

It is perhaps easiest to begin by thinking about this in two dimensions as, for example, in packing together coins which have been laid flat on a table. Figure 6.1 shows three two-dimensional structures formed by packing nine coins together. In the first, the basic arrangement is square but in going from Figure 6.1b to 6.1c the original square is progressively tilted; in doing this each of the four-sided spaces, representing the uncovered table surface showing between the coins, becomes elongated (Figure

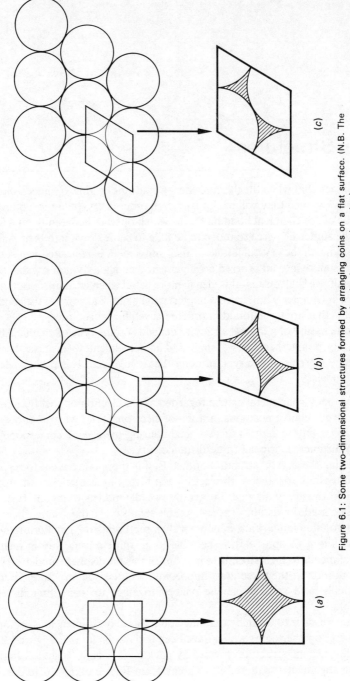

Figure 6.1: Some two-dimensional structures formed by arranging coins on a flat surface. (N.B. The shaded regions represent the area of uncovered surface showing between the coins.)

(a)

(b)

(c)

6.1*b*) and finally splits into two three-sided spaces (Figure 6.1*c*). The shaded regions in the lower half of the figure show that the area of uncovered surface progressively diminishes as the structure tilts more and more; to put this another way, we can say that the efficiency with which the coins cover the surface of the table progressively increases—they become more closely packed. It is not difficult to prove, with simple geometry, that the square packed arrangement (Figure 6.1*a*) results in 79% of the total surface of the table being covered whereas the arrangement in Figure 6.1*c* gives 91%. In fact Figure 6.1*c* represents the closest packed arrangement possible; the coins in the top left and bottom right hand corners of the structure have come into contact with the central coin and this prevents further tilting. The central coin now has six neighbours in direct contact rather than four (as in Figure 6.1*a* and 6.1*b*) and the figure shows that it is not possible to fit any more around its circumference; so six is the maximum possible number of neighbours in direct contact with any particular coin in an arrangement of this kind where all the coins have the same diameter.

We can now see that, in this two-dimensional model of a metal crystal structure, the closest packed arrangement occurs when each coin is surrounded by the maximum possible number of neighbours. But we already know enough about the metallic bond to be able to predict that this is the arrangement that we would expect if the coins in the analogy are replaced by metal atoms. Firstly, we know that the metallic bond is non-specific and non-directional; as far as any particular atom in a metal crystal is concerned, one neighbour is as good as any other irrespective of direction. Furthermore, we saw in Chapter 4 that when two atoms are close to one another, but not actually in contact, there will be a net attractive force which will tend to pull them together until they reach their equilibrium distance. This means that, in the arrangement of atoms shown in Figure 6.2*a* (which corresponds to the structure in Figure 6.1*b*), there will be an attractive force between atoms X and Y which tends to pull them into contact to give the close packed arrangement in Figure 6.2*b* (corresponding to Figure 6.1*c*). The structure in Figure 6.2*a* therefore has a relatively

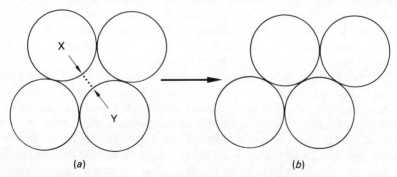

(a)   (b)

Figure 6.2: The attractive force between atoms X and Y will tend to make the structure collapse into the close-packed arrangement.

high potential energy which can be reduced by adopting the stable close packed structure in Figure 6.2*b*. Of course, the square packed arrangement of atoms in Figure 6.3 (corresponding to Figure 6.1*a*) will have an even higher potential energy than the structure in Figure 6.2*a*, but the tendency to tilt to the right will be exactly balanced by the tendency to tilt to the left and the upper two atoms will be perched on the lower two in a highly unstable way; in fact this structure would not normally exist because even a gentle push, provided by thermal vibration for instance, would be enough to upset the balance and cause it to collapse either one way or the other.

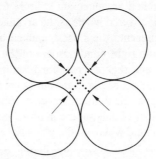

Figure 6.3: The tendency for the square packed structure to collapse to the right is exactly balanced by the tendency to collapse to the left.

So now we have a simple two-dimensional picture of metals in terms of the non-directional and non-specific nature of the metallic bond; we can see that the most stable arrangement exists when the atoms are close-packed and each atom has the maximum number of neighbours possible. But similar arguments apply to three-dimensional structures too; again, all the atoms tend to surround themselves with as many neighbours as possible so that their total potential energy is at a minimum.

We now need to examine what geometrical arrangements will result when we pack atoms closely together in three-dimensional structures. We already know that, in two dimensions, we can pack six neighbours around a central atom and we must now extend this idea to give a three-dimensional model. We can do this by regarding the basic arrangement in Figure 6.4*a* as a single layer of spherical atoms on top of which we shall build a second layer. Each atom in the figure is, for convenience, labelled **B** and between the atoms there are triangular hollows labelled either **A** or **C**. We can build up a second layer of atoms, on top of the **B** layer, by placing them so that they rest in these hollows. But it is not possible to simultaneously occupy all the six hollows shown in Figure 6.4*a*; we can either fill the three **A** hollows (Figure 6.4*b*) or the three **C** hollows (Figure 6.4*c*). We can see that the distance between two **A** hollows, or between two **C** hollows, exactly equals the diameter of one atom and Figure 6.4*d* shows that an upper layer built up by placing atoms in either the **A** or **C** hollows will itself be close-packed; by rubbing out the distinguishing letters, we find that there is no practical difference between Figures 6.4*b* and 6.4*c*—nevertheless there are two distinct sets of hollows between the **B** atoms and we could choose to build the upper layer in either of them. This distinction might appear to be trivial when we are talking about stacking only two close-packed layers of atoms together; but it becomes important when we come to think about adding a third.

In Figure 6.4*b* for instance, the layer of atoms in the **A** positions rests upon the atoms in the **B** layer; but in an extended three-dimensional structure the **B** atoms will

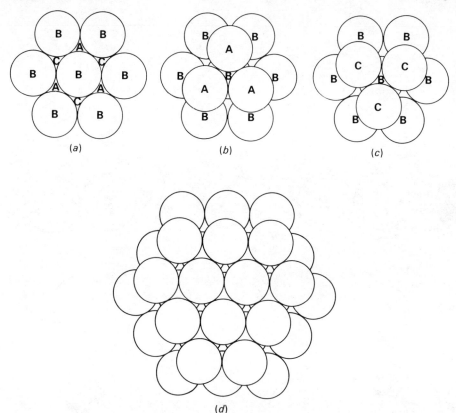

Figure 6.4: A second layer of atoms can be constructed on top of the close-packed layer in (a) so that it either occupies the A hollows as in (b), or the C hollows as in (c); (d) shows that the second layer will itself be close-packed as well.

themselves be supported by another layer below. And again there are, of course, two alternative positions for this lower layer; it can either fit into the underside of the **B** layer in the **A** positions (Figure 6.5a) or in the **C** positions (Figure 6.5b). This leads to two alternative types of crystal structure. In the first, the close-packed layers are stacked in alternating relative positions so that the arrangement throughout the crystal is ABABAB ... etc.; in the second, the relative positions are repeated every three layers, ABCABC ... etc.

The ABABAB stacked arrangement is called the *hexagonal close-packed* structure because, in extended three-dimensional form, it can be thought of as being constructed from many identical building blocks, or *unit cells* as they are called, which have a hexagonal (i.e. six-sided) form. Figure 6.6 illustrates how we can picture these unit cells and the way in which they fit together, like bricks, to form the hexagonal close-packed structure.

The ABCABC stacked arrangement is called the *face-centred cubic* structure. Figure 6.7 shows that, in this case, we think of the unit cell as a cube which has been constructed by stacking the close-packed planes perpendicular to the diagonal from

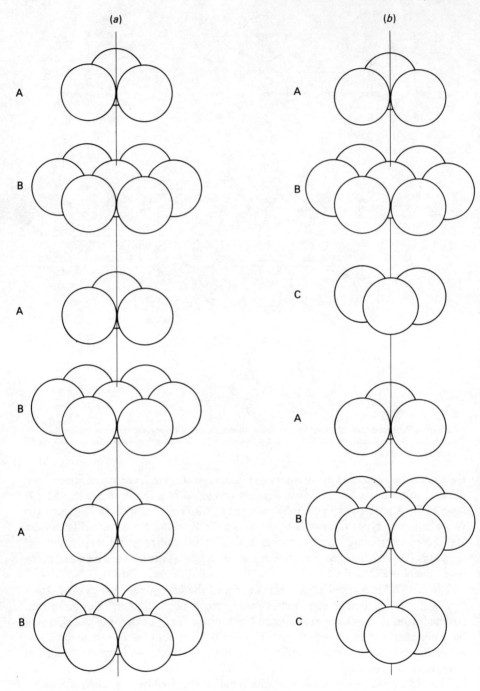

Figure 6.5: (*a*) ABABAB stacking of close-packed planes gives the hexagonal close-packed structure; (*b*) ABCABC stacking of close-packed planes gives the face-centred cubic structure.

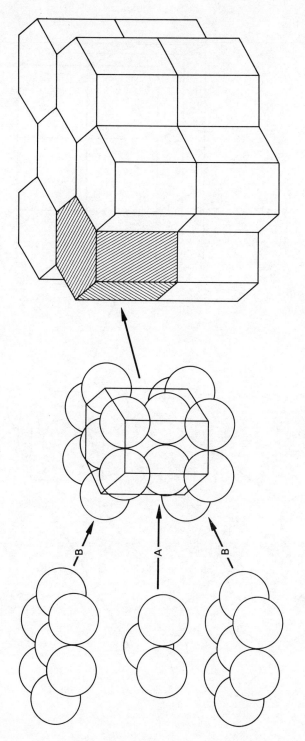

Figure 6.6: The hexagonal close-packed structure.

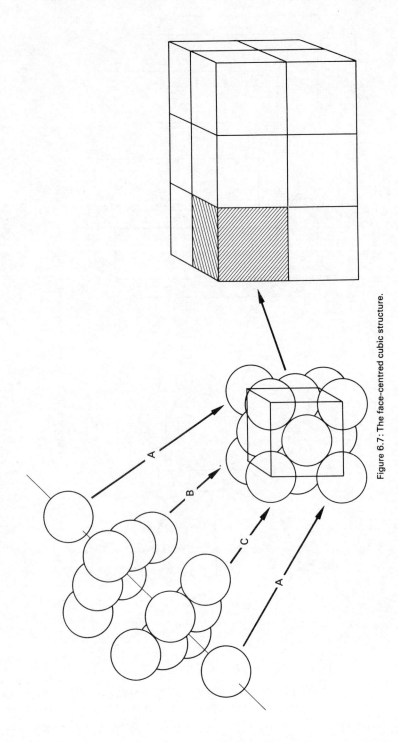

Figure 6.7: The face-centred cubic structure.

corner to corner through its centre; the figure shows that there is one atom at each corner of the cube and one at the centre of each face—hence the description *face-centred cubic*.

The importance of distinguishing between the two close-packed structures will become apparent in Chapter 8 when we discuss metals in more detail. But in the meantime we should note that, in both cases, every atom within the structure is surrounded by twelve neighbours in direct contact. As Figure 6.4*a* showed, any atom is surrounded by six neighbours in the same close-packed layer. Furthermore, Figures 6.4*b* and 6.4*c* show that this central atom will also be in contact with three atoms in the layer above; and, regardless of whether the stacking is ABABAB or ABCABC, there are also three in the layer below, making a total of twelve atoms in contact altogether. So we can regard each atom in these close-packed arrangements as being surrounded by a tightly packed shell of twelve neighbouring atoms; adjacent atoms in the shell are in contact with each other and so twelve is the maximum number possible.

We could, of course, pack less than twelve neighbours around the central atom and, although they would still be in contact with it, they would be able to spread out to some extent around its surface and lose contact with one another. This occurs in the *body-centred cubic* structure which is also adopted by some metals. Figure 6.8 shows that the unit cell is again a cube with an atom centred at each corner, but in this case there is also one atom at the body-centre of the of the cube. The body-centred atom is

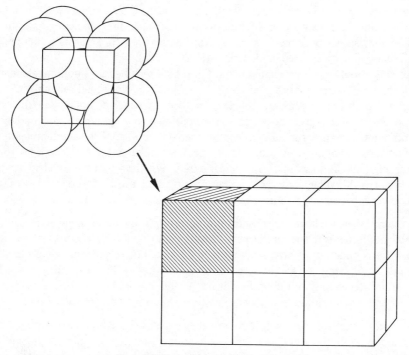

Figure 6.8: The body-centred cubic structure.

therefore in contact with only eight neighbours, which are the atoms at the corners of the cube; and adjacent corner atoms do not touch each other.

The body-centred cubic structure is therefore not so densely packed as the two close-packed structures that we considered previously. It is, in fact, quite easy to show this by using simple geometry to calculate the relative efficiency with which these arrangements can be used to pack spheres into a given volume. We shall not do the mathematics here but, if we did, we would find that the atoms in both the hexagonal close-packed and the face-centred cubic structures occupy 74% of the total volume of the crystal, the remaining 26% representing the empty space between the atoms. On the other hand we would find that, in the body-centred cubic structure, the atoms only occupy 68% of the volume of the crystal with 32% empty space; so by reducing the number of direct neighbours of each component atom to eight and spreading them out so that they no longer touch each other, the crystal structure becomes more open and less efficiently packed.

But why should any metal adopt the body-centred cubic structure? After all, the two close-packed structures with twelve neighbours around each atom should be more stable and we might therefore expect all metals to exist in either of these forms. And yet we find that the alkali metals (i.e. the Group I metals) and a number of transition metals normally adopt the body-centred cubic structure.

To take the alkali metals first, we should remember that they are soft and have low melting points; these are both consequences of the relatively low strength of the metallic bond in these elements—in other words, the low cohesive forces can fairly easily be overcome by mechanical stress and by heat. These rather weak bonding forces are unable to restrain the thermal vibration of the atoms sufficiently for them to pack in the closest possible way; the alkali metals therefore normally adopt the relatively open body-centred cubic structure.

But why should there be relatively hard and high melting transition metals which adopt the body-centred cubic structure? To answer this we should remember, from Chapter 3, that transition metals possess inner orbitals which do not contain their full complement of electrons. This suggests the possibility that, although the bonding in transition metals is predominantly metallic, there is some scope for overlapping of incomplete orbitals of adjacent atoms in the metal crystal and therefore scope for some partial covalent character in the metallic bonding. We know that the covalent bond is directional in character; it therefore seems possible that directional constraints of this kind prevent these metal atoms from adopting the close-packed structures.

To summarise, we have seen that the non-directional and non-specific nature of the metallic bonds tends to encourage metals to adopt close-packed structures where each atom is surrounded by twelve neighbours in direct contact. But some metals tend to adopt the less closely packed body-centred cubic structure with eight neighbours for each atom; depending on the particular metal, this is possibly due to the disruptive effect of thermal vibration or to partial covalent character in the metallic bond.

We can now move on to consider crystal structures based on ionic bonding. Like the metallic bond, the ionic bond is non-directional and non-specific and, again, we can imagine building up structures in terms of packing spheres together as closely as possible. But in this case we must think of the spheres as being electrically charged; in ionic structures there will be attractive forces between oppositely charged ions and repulsive forces between similarly charged ions.

On these grounds we would expect the two-dimensional structure in Figure 6.9*a* to be unstable; the negative ions would of course tend to be drawn into contact with the central positive ion but, in doing this, they would come into contact with each other and the resulting repulsive forces would make the structure unstable. By reducing the number of surrounding negative ions, as in Figure 6.9*b*, they can now come into contact with the central positive ion without touching each other—the structure is then

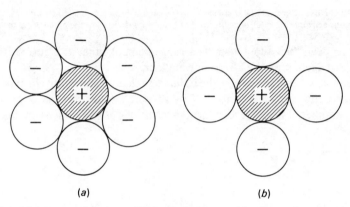

(a)   (b)

Figure 6.9: In (a) the arrangement is unstable because of the repulsive forces where the negative ions touch each other; in (b) the arrangement is stable because the number of negative ions has been reduced and they no longer touch each other.

stable. This argument would tend to suggest that ions are not able to adopt the highly close-packed structures that many metals can; we can think of ions as tending to surround themselves with as many oppositely charged neighbours as possible, but the extent to which they do so is limited by the condition that similarly charged ions should not come into contact with each other.

But this picture is oversimplified because the oppositely charged ions are generally of different sizes; Chapter 4 showed us that negative ions tend to be relatively large whereas positive ions tend to be small. The relative size of the ions has an important effect on the crystal structures which are formed by packing them together; a convenient measure of this relative size is the ratio of the radius of the smaller ion to that of the larger, and we can begin to look at the effect of this by using the two-dimensional analogy in Figure 6.10.

In Figure 6.10*a* the smaller central ion is large enough to hold the outer ions apart and the structure is therefore stable. If we imagine what would happen if we could successively reduce the size of the central ion we find a limit to the stability of this type of structure. This occurs when the central ion is just small enough to allow the outer ions to touch one another, as in Figure 6.10*b*; since the outer ions are similarly charged (probably negatively in this case, being the larger), there are strong repulsive forces where they touch and the structure becomes unstable—by removing one outer

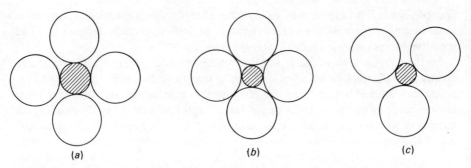

Figure 6.10: The arrangement in (a) is stable; in (b) the central ion is too small to prevent the outer ions from touching each other so, in (c), one outer ion has been removed to allow the others room to separate.

ion the others can then adopt a stable arrangement, like that in Figure 6.10c, where they are once again separated from one another.

So we can now begin to see how the radius ratio has an important effect on the type of crystal structure which is formed. If we assume that the number of positive ions equals the number of negative ions then, in terms of our two-dimensional model, the

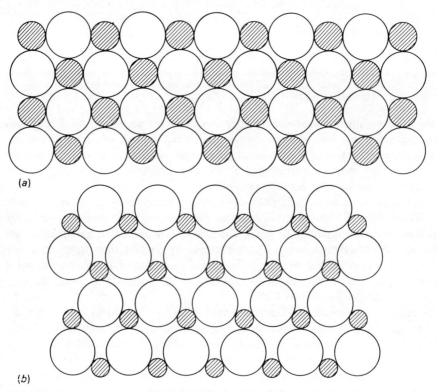

Figure 6.11: Two-dimensional models of ionic crystal structures: (a) corresponds to the arrangement in Figure 6.10a and (b) corresponds to the arrangement in Figure 6.10c.

arrangement in Figure 6.10*a* would lead to the extended structure in Figure 6.11*a*; each small ion is surrounded by four large ions in a square packed arrangement, and each large ion by four small ones. The arrangement in Figure 6.10*c* would lead to the structure in Figure 6.11*b* where each small ion is now surrounded by three large ones in a triangular arrangement, and each large ion by three small ones. (N.B. Where the size of the charge on the positive ion differs from that on the negative ion then, of course, the number of positive ions must differ from the number of negative ions for the crystal to remain electrically neutral; this will have an effect on the structural arrangement but, in this book, we shall only concern ourselves with specific structures which contain opposite ions of equal charge size and, hence, equal numbers.)

Although we shall not examine ionic structures in great depth, we should briefly extend our simple two-dimensional model a little way to look at the effect of radius ratio in real three-dimensional crystals.

As a starting point we can begin by supposing that we have a radius ratio of 1.00, i.e. that the positive and negative ions are of equal radius; so, on the basis of size considerations alone, we would expect to be able to fit twelve negative ions round the positive ion rather like a close-packed metal—but of course this would mean that adjacent negative ions would touch each other and therefore that the structure would be unstable on electrostatic grounds. However, we can still build a stable ionic structure from oppositely charged ions of equal size if we base it on the body-centred cube; Figure 6.12*a* shows that if we place the positive ion at the body centre we can then surround it with eight negative ions, centred at the corners, which do not touch each other. And Figures 6.12*b* and 6.12*c* show that, in building up the extended three-dimensional structure with equal numbers of oppositely charged ions, each negative ion is itself surrounded by eight positive ions.

But what happens when we reduce the radius ratio? In a similar way to our two-dimensional model in Figure 6.10, we find that the outer ions move in closer together as the central ion becomes smaller; eventually, at the limiting value of the radius ratio, they touch one another and the structure becomes unstable. Rearrangement is then necessary so that the central ion can be surrounded by a smaller number of outer ions which no longer touch each other.

Figure 6.13 shows successive stages in this process of structural rearrangement as the radius ratio is reduced from 1.00. Although we shall not do the calculations here, it is only a matter of simple geometry to work out the limiting radius ratio for each structure shown. For instance, in Figure 6.13*a* which corresponds to the structure in Figure 6.12, we would find that the outer ions come into contact with each other when the radius ratio is reduced to 0.73 (i.e. when the radius of the central ion falls to 73% of the radius of the outer ions). The structure is then readjusted so that the number of outer ions falls to six and they can then separate again, as in Figure 6.13*b*. As we reduce the size of the central ion further, the outer ions continue to move closer together until they touch again when the radius of the central ion has fallen to 41% of that of the outer ions (i.e. at the radius ratio of 0.41); the structure is then readjusted and the number of outer ions falls to four, as in Figure 6.13*c*.

The point of all this is that, in theory at any rate, we should be able to predict the

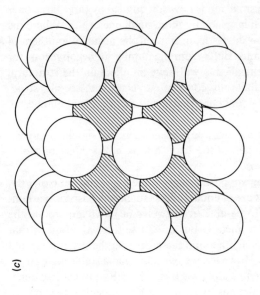

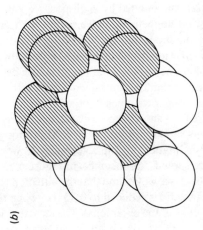

(c)

(a)

(b)

Figure 6.12: (a) Each positive ion is surrounded by eight negative ions and (b) each negative ion (represented by the top right-hand atom at the back of the original cube) is surrounded by eight positive ions to give (c), which represents the extended three-dimensional arrangement called the caesium chloride structure.

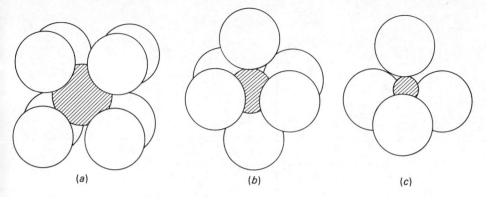

(a)   (b)   (c)

Figure 6.13: If the relative size of the central ion is successively reduced then the number of oppositely charged ions surrounding it must be correspondingly reduced as each limiting value of the radius ratio is reached.

structural arrangement that an ionic material will adopt if we know the size of the ions involved.

Taking caesium chloride as an example, the radius of the caesium ion (Cs^+) is 1.65×10^{-10} metre and the radius of the chloride ion (Cl^-) is 1.81×10^{-10} metre; and these values give a radius ratio of 0.91. But we know that the structure in Figure 6.13*a* is stable for radius ratios from 0.73 up to 1.00, so we should therefore expect this to be the arrangement that caesium chloride would adopt; this, in fact, is the case and we find that each caesium ion is surrounded by eight chloride ions and vice versa—and the extended structure of this general form, which is named after caesium chloride, is the one on which Figure 6.12 is based.

Taking sodium chloride as another example, the sodium ion (Na^+) has a radius of 0.98×10^{-10} metre, which is 54% of the radius of the chloride ion. We would therefore not expect to be able to pack eight chloride ions around each sodium because the limiting radius ratio for this arrangement is 0.73; but it should be possible to pack six, as in Figure 6.13*b*, because this is stable for radius ratios down to 0.41. So we should therefore expect sodium chloride to have a structure in which each ion is surrounded by six of the opposite charge. In fact, as Figure 6.14 illustrates, this is what we do find and this arrangement is called the rocksalt structure after the naturally occurring crystalline form of sodium chloride. Magnesium oxide has a radius ratio of 0.59 and therefore also adopts the rocksalt structure.

But now we come to a complication. Zinc sulphide forms crystal structures in which each zinc ion is surrounded by four sulphide ions, and vice versa. However, we also find that the radius ratio for zinc sulphide is 0.48 and, on these grounds, we would expect to be able to pack sulphide ions around each zinc ion as in the rocksalt structure. So zinc sulphide is restrained from adopting the more closely packed arrangement which we would ordinarily expect to be more stable. The reason for this seems to be that the Zn—S bond is not purely ionic but contains sufficient covalent character to impose directional constraints on the structure which is formed.

So the radius ratio can give an indication of the structure which a given ionic

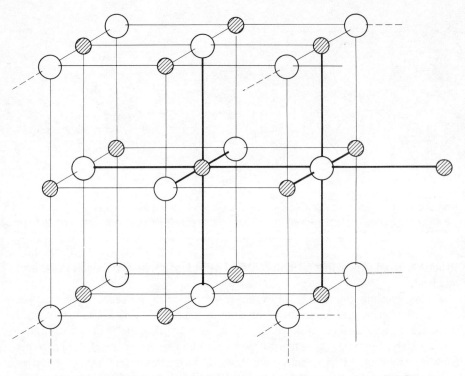

Figure 6.14: In the rocksalt structure each ion is surrounded by six of the opposite charge. (N.B. The structure is shown in expanded form for clarity.)

material will adopt, provided that the bond does not have sufficient covalent character to impose directional constraints.

This now leads us on to think about crystal structures which are held together by bonds which are predominantly covalent. We already know that the covalent bond has a highly directional and specific nature so, when we think about building up covalent crystal structures, we should really abandon the idea of packing spheres together as closely as possible. For instance, in Chapter 4 we met the idea that, in diamond, each carbon atom is surrounded tetrahedrally by four neighbouring carbon atoms; and Figure 6.15a indicates how this leads to an extended three-dimensional structure. Because each carbon atom is only surrounded by four neighbours then this structure has a very open form; but the high strength of the covalent bond, and the fact that each atom is firmly held in position by its neighbours, makes diamond a very strong and rigid material.

Pure carbon can also exist in the form of graphite, which is quite different from diamond; for instance, graphite is a good conductor of electricity and, furthermore, it is opaque—it seems to show metallic characteristics even though we normally regard carbon as a non-metal. To understand this we must look at the crystal structure of graphite in some detail.

Figure 6.15b shows that the carbon atoms in graphite arrange themselves in

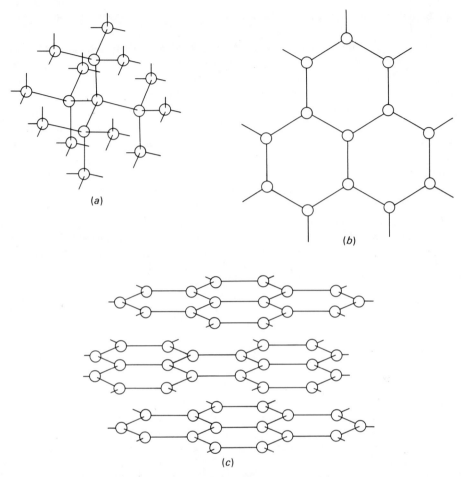

Figure 6.15: The crystalline forms of carbon: (a) the diamond structure; (b) the hexagonal arrangement within a graphite plane; (c) the graphite structure.

hexagonal patterns to form two-dimensional planes where each atom is covalently bonded to three neighbours; but this only involves three of the four outer electrons. In fact, the fourth electron on each carbon atom becomes *delocalised* to form a special kind of orbital which extends above and below the entire plane, rather like a sandwich; so the delocalised electrons are free to move within the plane and, in effect, can behave rather like the free electrons in a metal—and this gives graphite its metallic characteristics.

But so far we have only thought of graphite in terms of two-dimensional planes of carbon atoms; in the three-dimensional crystal structure these are stacked on top of one another as shown in Figure 6.15c. But, of course, all the outer electrons of the carbon atoms are now accounted for and there are none left to form primary bonds to hold the planes together; the necessary cohesion therefore relies on secondary forces of the van der Waals type. The planes are not very strongly bonded together so it is quite easy for them to slide over one another—and this gives graphite useful lubricating properties.

Graphite therefore relies on a combination of covalent bonding and van der Waals bonding. This leads us on to think about the general significance of secondary bonding forces in the structure of engineering materials.

In Chapter 4 we saw that the van der Waals bond, like the metallic and ionic bonds, is non-specific and non-directional; and this is illustrated by the fact that, when the inert gases are solidified, they adopt close-packed crystal structures. But secondary bonds like the hydrogen bond, which arise from permanent polarisation effects, can lead to directional constraints because the dipoles from which they originate are themselves directional; for instance, when two dipoles are close together, the attractive force between the positive end of one and the negative end of the other will tend to orientate them accordingly. Directional behaviour of this kind can have an important effect on crystal structure and a good illustration of this is given by ice.

We recall, from Figure 4.7 in Chapter 4, that the water molecule can be regarded as having two orbitals positively biased and two negatively biased. As Figure 6.16

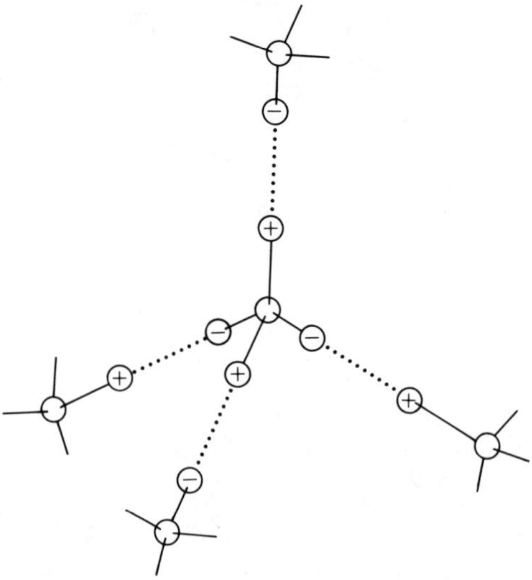

Figure 6.16: Hydrogen bonding in ice.

shows, this molecule can form a hydrogen bond with each of four neighbouring molecules; each of its orbitals attracts an oppositely biased orbital from a neighbour—and this is the basis of the crystal structure of ice, where we find each water molecule surrounded by four others in this way. But, as our earlier discussion of diamond would suggest, the fact that each molecule is only surrounded by four neighbours gives ice a very open structure. If the ice is melted, the rigid hydrogen bonded structure is disrupted and the molecules, no longer constrained in this open arrangement, can move closer together; so, in the liquid state, the molecules occupy less volume and water is therefore more dense than ice. This gives us a simple explanation of why water expands, and therefore bursts pipes, when it freezes and also, of course, why ice floats in water.

In Chapter 5 we saw that the low strength of the secondary bonds means that materials which are primarily dependent upon them for their cohesive strength usually have low melting and boiling points. As a result of this, these materials are generally of little value in constructional engineering. But secondary forces can play an important subsidiary role in the structure of some engineering materials. For instance, in Chapter 10, we shall see that plastics consist of extremely long chain-like molecules based on covalently bonded carbon atoms and that the secondary bonding forces between adjacent chains can have an important effect on their properties.

To summarise, this chapter has shown us how extended crystalline structures are formed under the influence of the cohesive forces of chemical bonding. We have seen that the non-directional and non-specific nature of the metallic bond leads us to think of metal crystal structures in terms of spherical atoms packed together as closely as possible; but we noted that thermal vibration can lead to a less densely packed structure—and also that a tendency towards covalent character in the metallic bond may lead to directional constraints which can have a similar effect. We have seen that ionic structures can also be thought of in terms of closely packing spheres together; but we had to take into account the repulsive forces between ions of like charge and then the effect of the relative sizes of the ions—and, again, we noted that covalent character in the bonding can lead to directional constraints. When we came to covalent structures themselves, we abandoned the idea of packing spheres together in view of the directional and specific nature of the covalent bond. Finally we noted that, although secondary bonding forces can be very important, they do tend to play a subsidiary role in the structure of engineering materials.

We are now in a position to be able to go on to consider how materials behave when we apply mechanical forces to them.

7.

Cohesive Forces and the
Mechanical Behaviour of Solids

We now have some understanding of how chemical bonding provides the cohesive forces necessary for the existence of the solid state. And we have seen that, in solid materials, atomic and molecular movement is restricted to thermal vibration about fixed equilibrium positions; and because they are held in their respective positions, it is very difficult for the component atoms (or ions or molecules) to move relative to one another—the material will therefore normally tend to maintain its shape. It is this tendency for solid materials to maintain their shape that enables them to resist mechanical forces; in this chapter we shall begin to use our knowledge of chemical bonding to develop an understanding of how they do this.

Firstly we will think about a simple experiment which demonstrates how materials behave under relatively small loads. If we suspend a length of wire by one end and hang a weight on the other then we find that the wire stretches; and, provided that we have not overloaded the wire and stretched it permanently, it will return to its original length if we remove the weight. If we carry out the experiment by increasing the load a little at a time, and carefully measure the corresponding increases in length, then we can plot a graph of load against elongation like those shown in Figure 7.1; these show that the greater the load, the further the wire stretches—and if the load is successively reduced then the wire will return to its original length along the load/elongation line. The figure also indicates that wires made from different materials stretch to different extents under the same loading conditions; the slope of the plotted line is a measure of the stiffness of the wire—a wire of low stiffness will stretch further under a given load than a wire of high stiffness.

So this experiment illustrates that, if we apply a load to a material, it responds by deforming; and the greater the load, the greater the deformation—furthermore, the deformation disappears when the load is removed provided that the material has not been overloaded and permanently distorted.

This behaviour provides the key to understanding how materials withstand mechanical forces. We shall see that the mechanical deformation of a material leads to the generation of internal forces within it which balance and support the externally applied force; the nature of these internal forces will particularly interest us because, as we shall see, they arise from the distortion of the chemical bonds within the material.

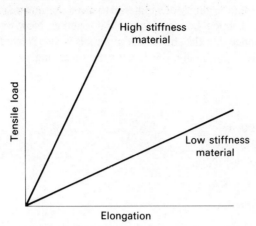

Figure 7.1: The elastic behaviour of materials under tensile load.

But before we go on to examine this idea we should note that, in our simple experiment, the measured stiffness of the wire will depend on its dimensions. For instance, it is harder to stretch a thick wire than a thin one of the same material. And the measured elongation under load depends upon the length of the wire; under the same loading conditions a piece of wire 2 metres long will stretch twice as far as if it was only 1 metre long. So we really need a true measure of the stiffness of the material itself which is not affected by the dimensions of the specimen; and to obtain this we normally plot stress* against strain* rather than plot the load applied to a specific specimen against the absolute change in its length. If we had plotted stress against strain in Figure 7.1 then the slope of the line, i.e. the stress required to produce a given strain, would have been a true measure of the stiffness of the material itself; in fact the slope of the stress/strain line is called the *modulus of elasticity*, or *Young's modulus,* of the material—and this type of behaviour, where a material shows a reversible response to stress in this way, is called *elastic behaviour.*

Now we can begin to look for an explanation of elastic behaviour in terms of chemical bonding forces. Figure 7.2 shows the melting points of the metals listed in Table 8.1 (on page 69) plotted against their stiffness, or Young's moduli. Although there is some scatter of the points, the figure quite clearly suggests that there is a general tendency for the stiffer metals to have higher melting points; the ability of a metal crystal structure to withstand the disruptive effect of thermal vibration seems to be related to its ability to withstand the distorting effect of mechanical loading. But in

---

* Some readers may be unfamiliar with the terms stress and strain.

If we define stress as being the load carried per unit cross-sectional area of the material, then this helps us to make direct comparisons between specimens of different dimensions under different loading conditions; for instance, if we hang a 2 kilogram weight from a wire having a cross-sectional area of 1 square millimetre then the stress in the wire will be the same as in a rod of 100 square millimetres cross-sectional area carrying a load of 200 kilograms.

Similarly strain can be a more useful way of expressing the resulting deformation; if we define strain as the ratio of the change in length under load to the original length then the actual length of the specimen will not affect the value of the strain measured in this way. For instance, if the wire was originally 1000 millimetres (i.e. 1 metre) long and it is stretched under stress by 1 millimetre to 1001 millimetres in length, then the strain is 1/1000 or 0.001; and this is the same as the strain in a wire 1 kilometre long which has been extended by 1 metre.

Chapter 5 we saw that the ability of a material to resist thermal disruption is related to the strength of the chemical bonds which hold it together. Now we can see that there appears to be evidence that the stiffness of materials is also related to the strength of the chemical bonds—and, really, this is not very surprising.

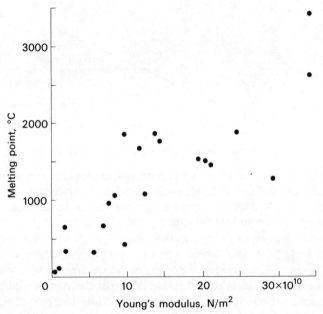

Figure 7.2: Melting point plotted against Young's modulus for some well-known metals.

It is perhaps easiest to look at the fundamental basis for the elastic behaviour of materials in terms of the net force/distance curve which we discussed in some detail in Chapter 4. The curve is shown again in Figure 7.3 and, for the purposes of our discussion here, we shall consider the general case of two unspecified atoms bonded together.

Under normal conditions the equilibrium distance between the two atoms is represented by the point Y in the figure; from Chapter 4 we remember that, at this point, a state of balance exists between the component attractive and repulsive forces which combine to give the net force/distance curve. But what happens if we try to pull the two atoms apart, for example to the point represented by Z? As the figure shows, a net attractive force will develop between the two atoms which will tend to pull them back to their equilibrium position. And if we try to squeeze the atoms together, to the point represented by X for example, a net repulsive force will tend to push them apart again.

This, then, is the basis of elastic behaviour and Figure 7.4 shows a schematic picture of the mechanical deformation of an *ideal* material in these terms. Figure 7.4*a* shows the material under tension and 7.4*b* under compression; Figure 7.4*c* shows the

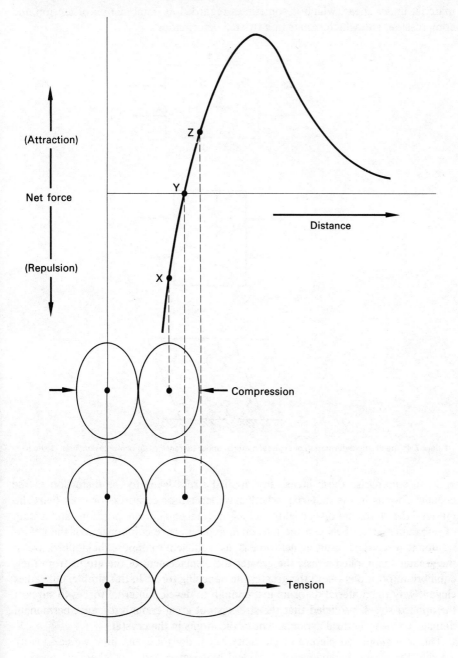

Figure 7.3: The elasticity of the chemical bond.

material under shear, which is sometimes regarded as combination of tension and compression, and which results in a twisted deformation.

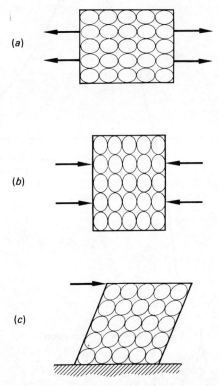

Figure 7.4: The elastic deformation of an *ideal* material under (a) tension, (b) compression, and (c) shear.

So to summarise these ideas, any applied stress leads to the distortion of the chemical bonds in the material; whether we stretch or compress or even twist the material the atoms are displaced from their normal equilibrium positions and a state of unbalance arises between the attractive and repulsive components of the net interatomic force. The resulting net force is just sufficient to oppose the applied stress; the greater the applied stress, the greater the displacement of the atoms from their equilibrium positions—and the greater the opposing force. In this simple, but rather elegant way the material deforms just enough to develop internal forces to support the applied stress, provided that the stress is not great enough to cause permanent damage to the structural arrangement of the atoms in the crystal.

This is a simplified picture of elasticity but it does indicate how we can justify viewing the elastic behaviour of our ideal material in terms of chemical bonding forces; and, as well as wires supporting weights, this simple model represents the fundamental mechanism whereby any solid object supports an applied load.

The slope of the interatomic force/distance curve, as it passes through the point Y, is of course a measure of the stiffness of the chemical bond itself—and this in turn determines the elastic modulus of the bulk material.

Figure 7.3 shows that, for relatively small displacements from the equilibrium position, the force/distance graph can be regarded as a straight line for practical purposes; in other words, provided that we do not deform the material too far, the amount of deformation is very nearly proportional to the applied stress which produces it*.

But the figure also suggests that, if we apply relatively high stresses to the material then this straight line relationship between stress and strain no longer holds, even approximately. For instance, if a progressively increasing tensile stress is applied then, in moving up the curve above the point Z, the slope decreases more and more rapidly although the deformation is still reversible if the stress is removed; in effect the stiffness of the material decreases at high elongations. Conversely, the slope of the force/distance curve increases under high compressive stresses; so, in effect, under these conditions the stiffness of the material increases.

We can extend these ideas to predict what we should expect to happen to the stiffness of the material as we raise its temperature. In Chapter 5 we used the two atom model to demonstrate thermal expansion; we found that the greater thermal vibration at high temperatures results in an increase in the equilibrium interatomic distance. So, in terms of Figure 7.3, an increase in temperature has the effect of shifting the equilibrium position up the curve from the point Y—and, in the light of the ideas that we have just been discussing above, we would expect this to lead to a decrease in the stiffness, or modulus of elasticity of the material. So our two atom model suggests that, if we raise the temperature of our material, its stiffness should decrease as it expands. And, in fact, the stiffness of most materials is found to decrease as the temperature is raised.

But we might also expect Figure 7.3 to be able to provide information about the strength of the material. After all, by increasing the stress on the material, we should eventually reach the peak of the force/distance curve; and this peak represents the force needed to pull the two atoms apart, i.e. the theoretical tensile strength of the chemical bond—so we should expect to be able to calculate a value for the tensile strength of the material quite easily.

However, in practice, the actual strength of real materials is generally very much lower than their calculated strength; and usually they break or are permanently deformed even before they can reach stresses corresponding to the significantly curved portion of the force-distance graph. Furthermore, our two atom model does not suggest to us how we might explain the different ways in which different materials fail under stress; for instance it does not explain why some materials are brittle but others are tough. Obviously our model needs some refinement.

However, before we go on to do this we should give some thought to how real materials behave—then we can define the shortcomings of the model.

We normally say that materials like glass and china are brittle because they tend to

---

* Readers who have previously studied elasticity will recognise this as the basis of Hooke's Law, which states that the extension of a material due to an applied force is proportional to the magnitude of the force.

be fragile and easily shattered into pieces. But the word *brittle* is used rather more precisely in a scientific context; and the clue to this is that, when we have broken a brittle material, we can reassemble the pieces like a jigsaw puzzle—if we glue them back together we find that the repaired object has almost exactly the same shape and size as it did before it was broken.

It is not difficult to build up a simple picture to explain this fracture process. If we apply a stress to the material it responds by deforming elastically; as the stress is increased the chemical bonds are distorted more and more until eventually they can support the stress no longer and the material breaks apart. But now the bonds within the broken pieces are no longer subjected to the applied stress so they return to their equilibrium positions—we can therefore glue the pieces back together again to reconstruct the original object.

So, at first glance, our simple theory of fracture seems to work for brittle materials like glass and china. But we meet a problem as soon as we measure the tensile strength of the material and check this against our calculated value based on the force necessary to pull two atoms apart. We generally find that the actual strength is very much lower than the theoretical value; the tensile strength of ordinary glass, for example, is normally about a hundred times less than theory would suggest.

But our simple model has other serious drawbacks. It demonstrates in principle why materials should be brittle but it does not explain why others are tough—for instance why metals do not normally shatter like glass. If we hit a piece of copper hard enough with a hammer we shall dent it permanently, or if we stretch it far enough it will remain stretched. And there are many other materials which can be deformed well beyond the point at which they can return to their original shape and size by elastic recovery. This is not to say that these materials do not eventually fracture because of course they do; but the fracture process is accompanied by permanent deformation so that we cannot fit the pieces back together to form the shape and size of the original object.

Permanent deformation of this kind is called *plastic deformation*—and it is important for us to clearly distinguish between *elastic deformation*, which is recovered when the stress is removed, and plastic deformation which is not. We can perhaps do this most easily by returning to our imaginary experiment in which we hung weights from a wire. Figure 7.5 represents the general form of the stress/strain curve that we could obtain in our experiment if we used a ductile metal wire, such as copper for example.

The region AB represents the elastic behaviour of the material; provided that we do not apply a stress greater than that at B then we can return to the point A by removing the load. But if the stress is increased above B we find that a marked curvature develops in the graph and its slope decreases; however this is not due to the shape of the interatomic force/distance curve because the deformation is no longer elastic. If, for instance, we remove the load at C the material will no longer follow the curve back along BA; instead it shows elastic recovery along the new line CA' and the distance between A and A' represents a permanent increase in the length of the wire. If we increase the stress on the wire again then it will now show elastic deformation along A'C and, if we continue to increase the stress after we reach C, then further perma-

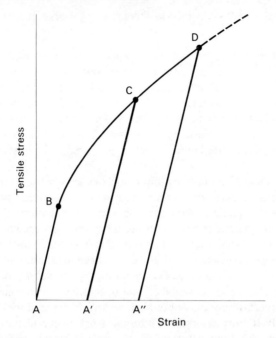

Figure 7.5: The stress/strain curve for a wire under tension showing elastic deformation, plastic deformation and strain-hardening.

nent stretching will continue as we proceed from C toward D. If we unload the wire at D then it will show elastic recovery along DA″, and this time the permanent increase in the length of the wire, represented by the distance between A and A″, is now greater.

We can see that plastic deformation has had the effect of increasing the maximum stress up to which the material can behave elastically. Before the wire had been permanently stretched it only showed elastic behaviour up to the point B. But the successive stretching operations had the effect of raising the limit of elastic behaviour firstly up to stresses corresponding to the point C and then to D. This strengthening of ductile metals by plastic deformation is called strain-hardening and we shall think about it again in the next chapter. We cannot of course continue stretching the wire indefinitely—sooner or later, depending on the material, it will break.

Our simple model has now run into further serious difficulties. As it stands it cannot explain how plastic deformation causes strain-hardening; and, indeed, it cannot even explain how plastic deformation occurs at all because, as long as the chemical bond remains unbroken, the two atoms will always return to the same equilibrium position when the stress is removed—the chemical bond cannot remain permanently stretched without a permanent force being applied to it.

And the problem can become even worse when we look at other materials. For instance, an ordinary rubber band can be stretched to several times its normal length by a relatively low stress; and when the stress is removed the rubber recovers elastically. But our simple two atom model would break apart long before such high elongations could be reached—and to do this, in theory, a very large applied force would be necessary.

So, initially at any rate, our simple model looked quite promising and we found that we could use it to represent normal elastic behaviour; we even used it to obtain a simple picture of brittle fracture, although it provided far too generous an estimate of the strength of ordinary brittle materials. But then it seemed to fall down rather badly; it failed to cope with two very commonly observed phenomena—namely, plastic deformation of metals and the elasticity of rubber. And we could find plenty of other examples where this idealised model would fail to predict the observed behaviour of real materials.

This does not mean that our two atom model in itself is wrong; our general view of the net force/distance curve does indeed represent the nature of the cohesive forces operating in solid materials. And there is some justification for viewing elastic behaviour, in these general terms, as the tendency for a material to resist an applied stress by means of internal restoring forces arising from the distortion of the chemical bonds. But what we have failed to do, so far, has been to take into account the overall internal structure of the material. For instance, in an analysis of the mechanical behaviour of a real three-dimensional crystal structure we must consider not only the nature of the force/distance curve but also the arrangement of the atoms (or ions or molecules) involved. Furthermore, the internal structure of real materials is normally far from perfect; thus, imperfections in the crystal lattice, and other defects introduced in manufacturing and handling, can all have an important effect on the behaviour of real materials.

So we must now look at real materials in rather more detail. And, as in our discussion of crystal structures in Chapter 6, we shall find it convenient to classify basic types of materials according to the nature of the chemical bonding forces which hold them together.

We shall begin by thinking about the nature of metals.

8.
The Nature of Metals

In Chapter 6 we saw that metal atoms in their crystal structures can be regarded as very small, hard spheres. We also examined some metallic crystal structures in terms of three-dimensional models based on various geometrical arrangements that can be formed when spheres are packed together. As we shall now see, these models can be very useful in discussing the behaviour of metals under stress.

At normal temperatures most common metals used in engineering adopt either the face-centred cubic, the hexagonal close-packed or the body-centred cubic structure (see Table 8.1). In fact, the majority of common metals are close packed (i.e. face-centred cubic or hexagonal close-packed). This is a consequence of the non-directional nature of the metallic bond; the bond pulls the metal atoms into the closest

FACE-CENTRED CUBIC	HEXAGONAL CLOSE-PACKED	BODY-CENTRED CUBIC
Aluminium	Beryllium	Sodium
Nickel	Magnesium	Potassium
Copper	Titanium	Vanadium
Silver	Cobalt	Chromium
Platinum	Zinc	Iron
Gold	Zirconium	Molybdenum
Lead	Cadmium	Tungsten

Table 8.1: The normal crystal structures of some well-known metals.

packing possible because there are neither the directional constraints of the covalent bond nor the electrostatic constraints of an ionic structure. The body-centred cubic metals, however, are not so tightly packed; they have a co-ordination number of eight (i.e. each atom has eight neighbours in direct contact) rather than twelve and the spheres occupy 68% of the total volume of the crystal rather than 74% as in the close-packed metals. In Chapter 6 we noted that, depending on the particular element, possible reasons for the relatively open structure of body-centred cubic metals might be either thermal vibration or some degree of covalency in the metallic bond.

How do all these ideas account for the way in which metals are deformed by applied stresses?

Firstly we must think about what can actually be seen when a metal crystal is permanently deformed. Figure 8.1*a* illustrates how failure can occur when a tensile stress is applied. The metal appears to become distorted by thin sections of the crystal slipping over each other. The crystal remains in one piece but, under the microscope, its surface is covered with fine parallel lines. In fact, the crystal appears to be deformed in much the same way as a pack of cards when a sideways force is applied to it (Figure 8.1*b*).

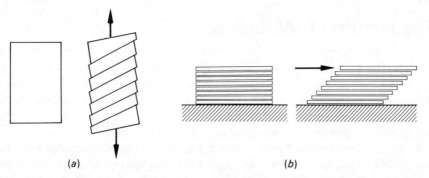

(a)   (b)

Figure 8.1: The analogy between the plastic deformation of (a) a metal crystal, and (b) a pack of cards.

Why this should happen can be better understood by first considering the two rows of atoms in the simple model shown in Figure 8.2. In the first case a tensile force is tending to split or cleave the top row away from the bottom and, in the second, a shear force is tending to make the top row slip over the bottom. It is fairly obvious that a greater force will be needed to completely separate the top row than merely to make it slip step by step along the bottom. In the first case the chemical bonding between the two rows must be completely overcome. In the second, the rows do not lose contact with one another; the force needed is only that to pull each upper sphere

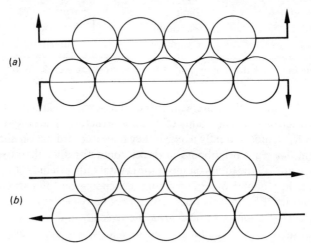

Figure 8.2: The relative movement of rows of atoms by (a) cleavage, and (b) slip.

out of the trough between two lower spheres and to make it slip over into the next trough along. In other words our model is weaker in shear than in tension.

Even though the crystal in Figure 8.1*a* is in tension it would appear to contain diagonal planes between which slip can occur in preference to cleavage at right angles to the applied tensile stress. The tensile stress is resolved into a shear stress across these diagonal planes and slip occurs between them.

We now need to understand how these *slip planes*, as they are called, can arise. Figure 8.3 represents a two-dimensional model of a metal crystal. Assuming an

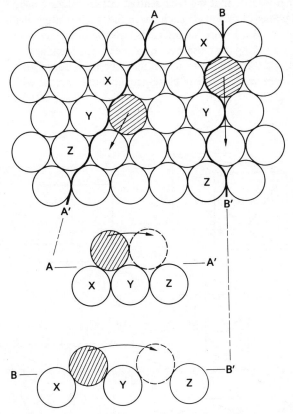

Figure 8.3: A two-dimensional model of a metal crystal showing how slip will occur more easily along close-packed directions.

applied stress is orientated similarly in each case we must decide whether slip would occur more readily between the rows of atoms separated by the line AA' or by BB'. To simplify the problem we can analyse it for each case in terms of the slip of the shaded atom from its rest position between the atoms X and Y to a new position between atoms Y and Z; the entire row containing the shaded atom will of course slip but it is easier to compare the alternatives in terms of the movement of one atom.

It is quite easy to see that in BB' the shaded atom lies in a deeper trough than in

AA′; furthermore, the distance between successive troughs is greater along BB′ than along AA′. For the shaded atom to move along BB′ more force is needed to pull it out of its trough and it must travel in longer steps than along AA′. Slip therefore occurs at a lower stress in the direction AA′. The atoms X, Y and Z are close-packed in the direction AA′ whereas they are relatively widely separated along BB′. The model therefore illustrates that slip will occur more readily in the more closely packed directions.

This idea can now be extended to three-dimensional crystals; to decide where slip is likely to occur it is necessary to identify the most closely packed planes.

In Chapter 6 we saw that the face-centred cubic structure is built up by stacking close-packed planes of spheres in such a way that their relative positions are repeated every three layers. But Figure 8.4 indicates how the same face-centred cube could be constructed by stacking the close-packed planes perpendicular to any one of the four diagonals running through the centre of the cube from corner to opposite corner; in other words, there are four sets of close-packed planes across which slip can readily

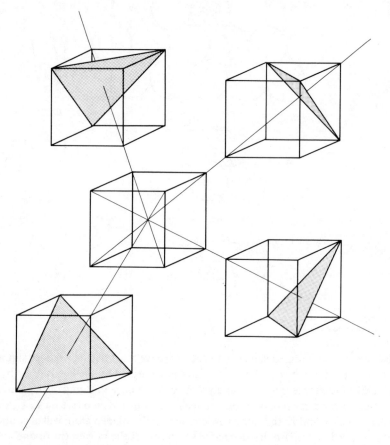

Figure 8.4: The four sets of slip planes in the face-centred cubic structure.

occur. Furthermore Figure 8.5 shows that there are three directions in which slip can readily occur across any close-packed plane. The atoms in the plane above can most easily move along the valleys between the close-packed rows; Figure 8.5 shows that there are three directions in which the rows are close-packed and therefore that there are three slip directions. Each slip direction in each slip plane constitutes a *slip system* and so there are twelve slip systems in the face-centred cubic structure.

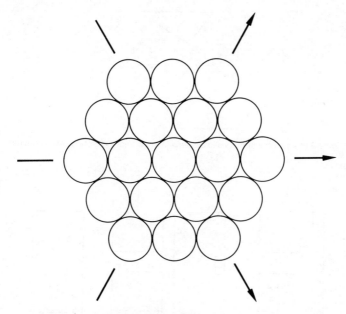

Figure 8.5: The three slip directions in a close-packed plane.

Whatever the orientation of a stress applied to the crystal there are always several slips systems available for plastic deformation. As a result of this, face-centred cubic metal crystals are highly ductile.

Chapter 6 showed that the hexagonal close-packed structure is also constructed from close-packed planes but that they are stacked so that their relative positions alternate. But the stacking of these planes in this way does not give rise to close-packed planes in other directions as we saw with the face-centred cubic structure; there is only one way of constructing the hexagonal close-packed structure. In this case there is therefore only one set of slip planes and, as shown in Figure 8.6, these lie parallel to the base of the unit cell. Again, there are three directions in which slip can readily occur across these planes; there is therefore a total of only three primary slip systems and these are confined to the single set of close-packed planes. If a stress is applied either parallel or perpendicular to the planes, as in Figure 8.7, then slip will tend not to occur and brittle fracture, without plastic deformation will result. On the other hand, if the close-packed planes are at some diagonal orientation to the applied stress then slip can occur.

In fact there are other, but less closely packed, planes across which slip can

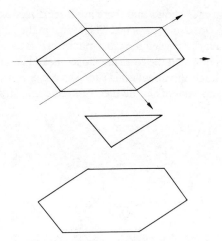

Figure 8.6: The three primary slip systems in the hexagonal close-packed structure.

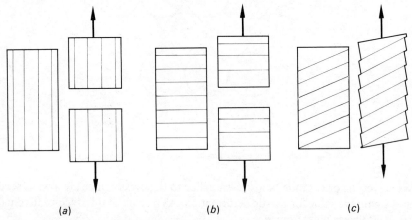

(a)   (b)   (c)

Figure 8.7: A tensile stress applied (a) parallel, or (b) perpendicular to the slip planes in the hexagonal close-packed structure will tend to give brittle fracture; in (c) the stress is applied at a diagonal orientation and slip can occur.

sometimes occur in hexagonal close-packed structures. However, in general, this type of metal crystal is less ductile than the face-centred cubic because the overall scope for slip is reduced.

Lastly we need to consider how the less densely packed body-centred cubic structure behaves. This has no close-packed planes. Nevertheless the most dense packing is found in the planes which pass through the diagonally opposite edges of the unit cell (Figure 8.8). Slip can occur along these and there are two slip directions in each. Figure 8.9 shows that there are six such planes making a total of twelve slip systems of this kind. There are other slip planes in the structure and, in all, forty-eight slip systems have been identified. But the body-centred cubic structure is less densely packed than the face-centred cubic (it has a co-ordination number of eight rather than twelve) and therefore tends to be less ductile; however it is generally more ductile than the hexagonal close-packed structure which, as we have seen, provides limited scope for slip.

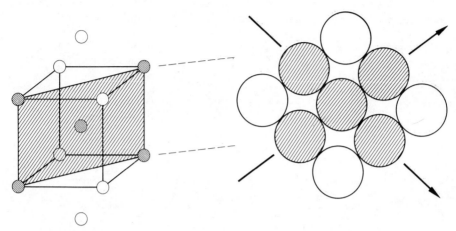

Figure 8.8: One of the most closely packed planes in the body-centred cubic structure showing the two slip directions.

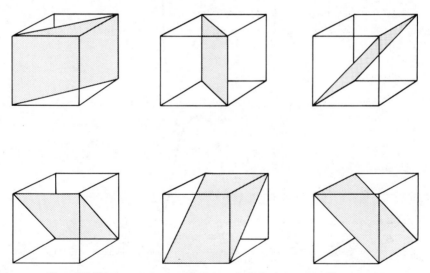

Figure 8.9: The six most closely packed planes in the body-centred cubic structure.

To summarise at this point, we now have a picture of the ductility of metals in terms of the slip mechanism. The geometry of the various crystal structures leads to preferred directions in which slip will tend to occur; different structures result in varying degrees of scope for slip. Based on this, we have been able to go some way in deciding which metals should be more ductile than others.

It is possible to extend our simple model of the slip mechanism to estimate how strong metals should be. Again we can simplify the argument by considering the slip of a single atom as illustrated by Figure 8.10. This time we shall look at how the sideways shear force varies as we pull the shaded atom from its trough between atoms X and Y to the trough between Y and Z. As the force/displacement curve

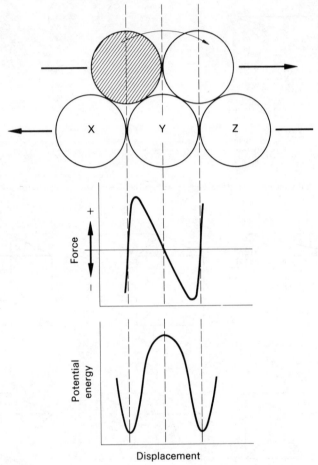

Figure 8.10: The variation of the applied force and the potential energy during one cycle of the slip process.

shows, the force needed must increase to get the atom to begin to climb up out of the trough between X and Y. A maximum is reached as the atom moves up the hill but as it begins to approach the top, and the slope of the hill decreases, the force needed will also decrease. When the atom reaches the point at which it is balanced directly on top of Y the force is zero. As the atom passes the crest of the hill it tends to run downwards into the trough between Y and Z; the force then becomes negative since, in effect, it would have to be reversed to stop the atom running down. When the atom reaches the rest position at the bottom of the trough the force again returns to zero. For slip to occur a further step along the plane the cycle is repeated.

In terms of this model, slip continues as long as the applied force exceeds the peak force needed to pull the atom out of each successive trough as it moves along the slip plane below. It is possible to estimate the value of the peak force and then to use this to estimate the stress that a metal crystal will withstand before plastic deformation occurs.

But before we go on to consider the result of this it is helpful to think about how the energy of the system varies as the slip process occurs. The two troughs in the model represent positions of minimum potential energy which are separated by an energy barrier; this is illustrated by the potential energy/displacement curve shown in Figure 8.10. Work must be done to pull the atom up to the top of the hill just as work must be done to propel a motor car up a hill. Now what happens to the stored potential energy when the atom runs down the other side? The clue to this is given by bending a piece of metal backwards and forwards. We notice that it gets hot. This leads us to the idea that the stored potential energy is converted to heat. In terms of our simple model, the atom does not simply roll into the trough between Y and Z and stop; it continues to run on up the other side. It will of course then slow down, stop and roll back again through the trough and back up towards the summit above the atom Y; in this way it will continue to oscillate about the rest position at the bottom of the trough. In terms of the motor car analogy in Chapter 1, the oscillations will fairly rapidly die down because of frictional losses between moving parts. For an atom in a crystal the oscillations will decrease since they will be transmitted to neighbouring atoms. These oscillations of course represent vibrations of the crystal lattice or, in other words, heat; the potential energy of the atom at the top of the hill is converted to heat energy and the temperature of the crystal will rise. In bending our piece of metal a considerable amount of slip, or plastic deformation occurs and the work done to achieve this is mostly dissipated as heat. So if we drop some metal object on the floor it is unlikely to break into pieces. What will tend to happen instead, if it falls hard enough, is plastic deformation in the region surrounding the point at which it actually touches the ground. The kinetic energy of the metal object at the moment of impact will be converted to potential energy of the metal atoms and will then be dissipated as heat.

We have now seen that our model should be able to provide an estimate of the strength of the metal crystal as well as give us a simple analogy to help us understand the toughness of metals.

But there is an important inconsistency. The estimated strength of a metal crystal, based on the model, differs very greatly from actual strengths measured in the laboratory; in fact the estimated strength turns out to be roughly a thousand times too great. Clearly there is some weakening mechanism in real crystals which we have not yet taken into account.

To understand this we should first think about the rate at which crystals grow. We know that atoms are extremely small; as we noted in Chapter 1, a straight row of four million copper atoms will only be just over a millimetre long. But crystals often grow by a millimetre in a few hours—sometimes considerably faster. The rate at which layers of atoms are deposited on the surface of a growing crystal is therefore often very high. It is then perhaps not surprising that these layers are not always perfect. Occasionally a layer may not have a chance to be completed before it is covered over by the next layer. Figure 8.11 shows an incomplete layer of this kind in a crystal lattice; this type of imperfection is observed in real crystals and is called a dislocation. The figure also shows how the lattice is distorted around the tip of the dislocation; the atoms around it are forced out of their stable, low potential energy positions.

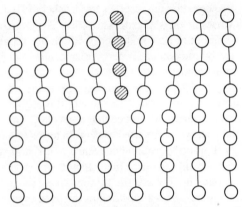

Figure 8.11: A dislocation in a crystal lattice. (N.B. The upright lines between the atoms merely serve to emphasise the layers and do not imply specific chemical bonds.)

The presence of dislocations can be used to explain why slip occurs in crystals at much lower stresses than might otherwise be expected. A popular analogy is to think about what happens if we try to slide a carpet across the floor. If we pull on the edge of the carpet there is considerable frictional resistance. On the other hand, if we form a wrinkle at one edge and push the wrinkle across the carpet we still move the carpet across the floor, but much more easily because we do it a little at a time. In Figure 8.12 this analogy is extended to a crystal structure; the dislocation is the equivalent of a wrinkle in a crystal. When a shear stress is applied to the crystal the dislocation tends to move. The next layer of atoms in its path will break adjacent to the tip of the dislocation. The lower half of the freshly broken plane will form a new complete plane with the original incomplete plane; the top half now becomes a new incomplete plane. Thus the dislocation has moved along one step. If the shear stress is maintained then the process is successively repeated and the dislocation runs through the crystal like a wrinkle across the carpet. The crystal is deformed a little at a time; the region behind the dislocation has slipped but the region in front awaits the arrival of the dislocation for slip to occur. The atoms in front of the dislocation will resist its movement because, as it approaches, it will tend to push them out of their stable, low energy positions. However, the atoms behind it will tend to run downhill into their stable positions and will tend to push the dislocation forwards. When no stress is applied to the crystal these effects oppose one another and the dislocation remains stationary. But the balance is easily upset and only a small shear stress is needed to cause the dislocation to move.

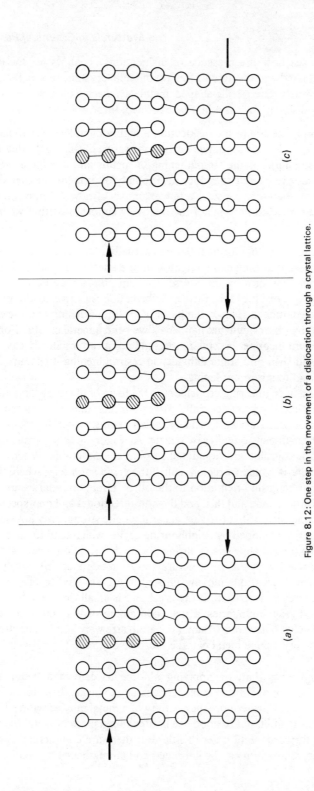

Figure 8.12: One step in the movement of a dislocation through a crystal lattice.

We can now see how the presence of dislocations accounts for the difference between the estimated and the actual strength of metal crystals. It is, in fact, possible to produce extremely fine needle-shaped metal *whiskers* which are crystals containing very few dislocations; as might be expected, these are very strong.

From this simple picture of the dislocation a highly complex science has grown, however a more detailed description is beyond the scope of this book. But an important aspect of metallurgy is the strengthening of metals and it is therefore often of great practical value to be able to reduce the mobility of dislocations to make metals stronger. We shall therefore consider a few examples of how the movement of dislocations can be restricted in terms of the relatively simple ideas that we have been discussing.

Firstly we should note that dislocations can actually multiply during plastic deformation; furthermore that they can interact with one another. If a metal is deformed sufficiently dislocations moving on intersecting slip planes may become entangled and interlocked; as a result of this, further deformation is restricted and the metal is hardened and strengthened. This is the basis of strain-hardening and the effect can be demonstrated by observing what happens if we bend a metal crystal. For instance we would find a small copper crystal, in the shape of a rod, relatively easy to bend; however we would then find some difficulty in restraightening it because the initial bending would lead to strain-hardening.

But metals used in engineering are seldom in the form of the single crystals that we have been discussing. They are generally *polycrystalline*; that is to say they take the form of a mass of small crystals, usually called *grains*, which interlock with one another like a three-dimensional jigsaw puzzle. As Figure 8.13 suggests, the crystal lattices of individual grains are orientated in different directions. Where one grain meets another there is a narrow region, called the *grain boundary*, where the transition occurs from one grain orientation to the other. Grain boundaries are normally only one or two atoms wide and this, and the non-directional and non-specific nature of the metallic bond means that they are generally quite strong. Any individual grain is therefore closely surrounded by neighbouring grains which tend to restrict the slip processes causing it to change shape during plastic deformation; because of the discontinuity in the regular crystalline structure, the grain boundary effectively provides a barrier at which dislocations moving through the grain tend to pile up—further deformation is therefore restricted. If the grain size is small there will be a relatively high proportion of grain boundaries; a metal containing small grains will then tend to resist continued deformation more than if the grain size were larger. The strength and ductility of a polycrystalline metal can therefore be controlled by adjustment of the grain size.

So we can now see that, by applying a force to deform a metal, there are mechanisms which can operate to make it necessary to increase the force for further deformation to occur. Furthermore the ductility of a metal may be reduced by plastic deformation. This is often put to good use in breaking a piece of metal by bending it backwards and forwards until it gets brittle. But there are commercial applications too; for example, aluminium can be hardened and strengthened by cold-rolling.

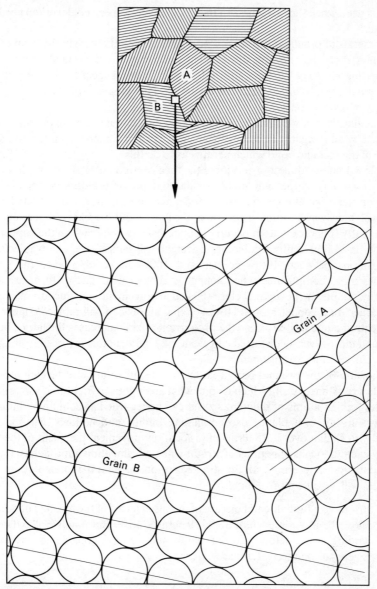

Figure 8.13: The polycrystalline structure of a metal showing an enlarged view of a grain boundary.

Clearly a metal which has been hardened in this way may be under considerable internal strain; many atoms will have been forced out of their most stable positions. If the temperature of the metal is raised the lattice vibrations will increase until the strained atoms can reorganise themselves into more stable positions without the metal, as a whole, melting; eventually recrystallisation will occur and the metal will return to its original *unworked* condition. This is the basis of the annealing process which is used when the brittleness of metals which have been cold worked needs to be reduced.

But metals in practical use are seldom pure; usually they are in the form of alloys containing more than one metallic element and sometimes non-metallic elements too. By inserting atoms of different sizes into crystalline arrangements of uniform atoms the original crystal lattice becomes distorted. This idea can be used to reduce the mobility of dislocations.

As an illustration we shall think about how the crystal structure of copper can be modified by alloying it with other metals. As we saw in Chapter 4 it is the non-specific nature of the metallic bond which permits us to do this.

Table 8.1 shows that copper adopts the face-centred cubic structure; but so does nickel. If we alloy copper with nickel we find that the nickel atoms can replace copper atoms on the original copper crystal lattice; because the atoms are of different sizes there will, however, be some distortion of the lattice. But the sizes of the atoms do not differ greatly (the nickel atom occupies about 8% less volume than the copper atom) and the basic face-centred cubic arrangement will persist over the whole range of compositions between 100% copper and 100% nickel. However the situation becomes more complicated when copper is alloyed with zinc to form brass. Zinc normally adopts the hexagonal close-packed arrangement and zinc atoms are about 13% larger than copper atoms. However zinc atoms can still replace copper on the face-centred cubic lattice but as the proportion of zinc increases the lattice becomes more and more severely distorted and eventually the structure is forced to change to a body-centred cubic arrangement.

Distortion of the original parent lattice by the substitution of foreign atoms can provide a resistance to slip which is somewhat analogous to friction. If a crystal plane includes foreign atoms of different sizes it can be regarded as containing bumps and hollows; an adjacent plane, also containing bumps and hollows, will have some difficulty in sliding over it. In terms of dislocation movement the misfit of the foreign atoms leads to localised regions of strain; when a dislocation approaches, the strain around its tip will interact with the strained regions around the foreign atoms and its progress will become more difficult.

We now have a fundamental picture of metals in terms of the metallic bond. Firstly we saw how the more important metal structures arise. But now we also have some understanding of the deformation of metals and we have gone some way in seeing how this depends on their internal structure.

9.

The Nature of Ceramics

The word *ceramics* means different things to different people; in the past it has tended to be confined to art-work made from materials like porcelain but nowadays it takes on a much broader meaning. Because of this, it is perhaps easier to begin by deciding what materials are obviously not classified as ceramics. Firstly, ceramics are non-metals; they do not rely on metallic bonding for their strength. On the other hand they are not organic; organic materials, like wood and plastics, have covalently bonded carbon as their basis and secondary bonding is nearly always an important feature as well.

This leaves a wide range of materials that we can broadly describe as ceramics. This includes bricks, pottery, inorganic glasses and enamels, cement and concrete, and even rocks and minerals. A common feature of all of these is that they depend on either ionic or covalent bonding, or a combination of both. The bonding electrons therefore tend to be localised and ceramics are generally relatively poor conductors of heat and electricity; in fact many are used as insulators. The bond strengths are greater than those of the secondary bonds in organic materials and often greater than those of the metallic bonds in metals; ceramics therefore usually have relatively high heat and chemical resistance.

We shall begin by thinking about the structure of ceramics.

Ceramics are often compounds formed between metallic and non-metallic elements, for example aluminium oxide. Because there is generally more than one element involved, ceramic crystal structures tend to be more complex than those of pure metals. They can be fairly simple, as in the case of magnesium oxide. But they are often quite complicated, as in natural silicates like mica and asbestos. Furthermore we find that some ceremics are non-crystalline and possess structures very similar to liquids; these are the materials that we call glasses. There are also composite ceramics, like the fired materials based on clay which contain both crystalline and glassy phases.

We have already met some of the simpler ceramic structure in Chapter 6. For instance we looked at the way in which sodium and chloride ions pack together to form the rocksalt structure; and we saw that magnesium oxide adopts the same arrangement. Another relatively simple structure is silicon carbide (SiC), commonly called carborundum, which is a covalent ceramic; we can regard this as being based on the diamond structure where each alternate carbon atom has been replaced by a silicon atom.

In silica (SiO_2) the Si—O bond is intermediate between pure ionic and pure covalent. Silica can exist in several crystalline forms (one of which is quartz) in which each silicon atom is surrounded tetrahedrally by four oxygen atoms and each oxygen links two silicon atoms. The result is a rigid three-dimensional structure containing twice as many oxygen atoms as silicon atoms. For our purposes we can conveniently represent crystalline silica by the two-dimensional model shown in Figure 9.1a. But Figure 9.1b shows that silica can also exist as a non-crystalline glass. This structure

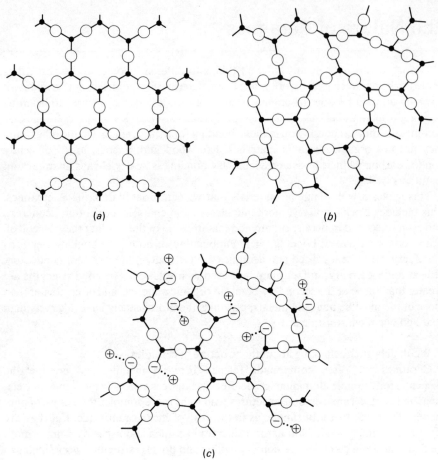

(a)

(b)

(c)

Figure 9.1: Two-dimensional models of various forms of silica:

(a) crystalline silica;
(b) silica glass;
(c) soda glass.

Key: ● silicon atom
            ○ oxygen atom
            ⊕ sodium ion

(N.B. In these two-dimensional models each silicon atom is, for convenience, represented as being sur-rounded by three oxygen atoms rather than by four as in the real three-dimensional structures.)

can be regarded as the form of silica which results when it is cooled too rapidly from the liquid state for recrystallisation to take place; in effect, the high strength of the Si—O bond means that the relative movement of the atoms in the liquid is very difficult—the viscosity is therefore very high and the ordering of the molten silica into a crystalline structure is restricted. Even though the cold glass is solid (i.e. relative movement can no longer occur) its structure is like a liquid because the long range repetitive order to crystalline silica is now lost.

Silica is the basic constituent of most commercial glass although various additives are used to modify its structure and properties. To help commercial processing, the high melting temperature and high melt viscosity of pure silica may be reduced by incorporating modifiers to partly break down the three-dimensional network. For instance in soda-glass, which is widely used in windows and bottles, the structure is modified by sodium oxide (Na_2O). Figure 9.1c shows the result of this; we can see that several Si—O—Si links have been broken and the broken ends *sealed* by an ionic bond between the terminal oxygen atom and a positive sodium ion. If sufficient breaks are made in the structure the glass becomes more mobile and workable at much lower temperatures.

The high strength of ionic and covalent bonds would lead us to suppose that ceramic materials should be intrinsically very strong. And we can perhaps argue that they might be expected to be as strong, if not stronger than most metals.

We have already seen that slip processes in metals tend to cause them to fail by plastic deformation at much lower stresses than would be required for direct bond breakage; but slip processes such as these are greatly restricted in ceramics. For instance, the covalent bond is strong, directional and specific between the two bonded atoms; this means that covalent crystal structures, such as diamond and silicon carbide, are highly resistant to the shear forces which cause slip in metallic crystals and that they are therefore very rigid and strong. In ionic crystals the slip processes are also restricted. This can be illustrated by considering the slip of one row of ions over the other in the simple model of an ionic structure shown in Figure 9.2. For the top plane to reach the next low energy position the negative ions in each row would have to come into contact with each other. This would lead to strong repulsive forces with the result that the two rows would tend to separate from one another. In other words, fracture will tend to occur by cleavage rather by slip; there will be no plastic deformation and the fracture will therefore be brittle.

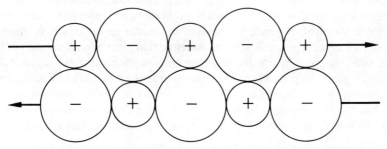

Figure 9.2: A simple model of an ionic structure under an applied shear stress.

Finally, in the case of glasses, there are no slip planes and this, and the high strength of the bonds, means that fracture will tend to be brittle and should require a high applied stress.

So we can see that, whereas slip tends to occur before fracture in metals, in ceramics the mechanism for slip is restricted by the nature of their chemical bonding and often by their structural features. We would therefore expect ceramics to be stronger than metals and also to be more brittle.

It is certainly true that ceramics generally break in a brittle manner with little or no plastic deformation; for instance, the pieces of a broken teacup or a broken window can be fitted back together like a jigsaw puzzle. But ceramics tend to break at lower stresses than the arguments above might suggest. Table 9.1 illustrates that their tensile strengths are often lower than those of metals. So we now have to look for some weakening mechanism in ceramics which makes them break at lower stresses than would be predicted on theoretical grounds.

Material	Tensile strength, N/m^2
Aluminum:	
(a) high purity	45×10^6
(b) commercial quality	$100-150 \times 10^6$
(cold worked)	
Copper (annealed)	220×10^6
Brasses (annealed)	$270-400 \times 10^6$
Magnesium oxide	70×10^6
(unpolished single crystal)	
Glass (ordinary)	$35-170 \times 10^6$
Brick	6×10^6
Concrete	5×10^6

Table 9.1: Some approximate tensile strengths for various metals and ceramics.

It is not too difficult to estimate the strength of a ceramic by calculating the force that should be required to pull two adjacent layers of atoms apart. On this basis, the theoretical strength of glass is roughly a hundred times the measured strength that is normally found. But if an ordinary glass rod is heated and drawn into a very fine fibre we find that its tensile strength is increased sometimes to values which closely approach the theoretical value. It appears therefore that glass contains some weakness which can be removed by the drawing operation. Modern theories suggest that this weakness is due to the presence of tiny flaws or cracks in the surface of normal glass; freshly drawn fibres contain very few surface defects and their strength is extremely high. This can be confirmed by touching the surface of a freshly drawn fibre; its strength is immediately reduced because even very gentle contact will result in surface damage.

Now how do these surface cracks actually reduce the strength of materials?

If we apply a tensile stress to a rod then, under ideal conditions, we can imagine lines of stress passing straight down the rod as in Figure 9.3*a*; these stress lines are simply a way of representing how the stress is carried through the rod, from one end to the other, by the chemical bonds. The effect of introducing a surface crack is shown in Figure 9.3*b*; the stress lines show how the load has to be carried around the tip of

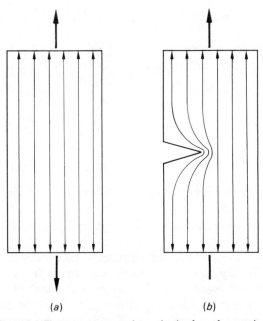

(a) (b)

Figure 9.3: The stress concentration at the tip of a surface crack.

the crack. In the diagram we see that the lines become very dense around the crack tip and that the chemical bonds in that region have to carry more than their fair share of the total load; in other words, stress concentration occurs at the crack tip. If the crack is longer then there are more lines to be diverted around its tip and the stress concentration is greater. In a very narrow crack the tip might conceivably be closed by only a small number of chemical bonds, or in the extreme case by just one as shown in Figure 9.4; in the figure this single bond has to carry a very much greater load than the average throughout the material. In fact, when a relatively low stress is applied to materials where even the strongest bonding forces are involved, the stress on the bond at the crack tip can be amplified so much that the bond fails. Once that bond fails the burden falls on the next; but now the crack is slightly larger and the stress concentration slightly greater. If the original load on the material is maintained then this bond will fail. So we can now see that cracks may grow, or propagate at relatively low stress levels in strong materials by the successive rupture of chemical bonds at the crack tip.

Since the strength of ceramics is so dependent on the presence of these cracks we must now decide what conditions are needed for them to propagate and for fracture to occur. As we shall now see, an energy balance is involved.

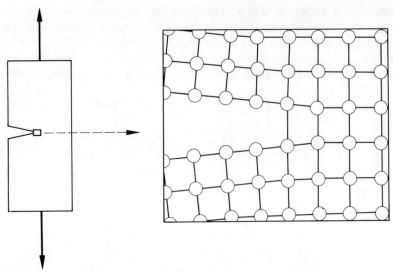

Figure 9.4: The bond at the tip of the crack in the stressed material is highly strained.

Firstly we must understand that work needs to be done to create new fracture surfaces.

In Chapter 5 we saw how the cohesive forces within a liquid make it tend to reduce its surface area to a minimum. The atoms or molecules at the surface of the liquid are only subjected to an inward force of attraction and therefore they possess higher energies than those in the interior. The liquid responds by attempting to minimise its surface energy and hence its surface area. Furthermore we noted that it is necessary to do work to increase the surface area of a liquid.

The same basic forces operate in a solid; the atoms in the surface of the material are only attracted inwards whereas those in the interior experience a more or less uniform field of attraction in all directions. The atoms at the surface therefore possess higher energies than those inside. It follows that energy must be provided to create the new surfaces in the fracture of a solid; in other words work must be done to propagate a crack. We must now identify the mechanism which is capable of doing this work.

Energy is stored in a stressed material. For instance energy is stored in the stretched rubber of a catapult and this can be used to propel a stone; the *strain energy* in the rubber is converted to kinetic energy of the stone as soon as it is released. But this strain energy could equally well be converted into electrical energy by using it to drive a small generator, or it could be converted into potential energy by using it to pull a weight up a slope. So strain energy in a stressed material can be converted to other forms of energy in a variety of ways. As we shall now see, it can be converted into surface energy and this, in fact, leads to a mechanism whereby cracks can be propagated through stressed materials.

Figure 9.5*a* represents a flat sheet of material which is not subjected to any stress; horizontal lines have been marked on its surface to enable us to see the strain in the material when a stress is applied. In Figure 9.5*b* a tensile stress is applied which leads to stretching of the chemical bonds; the horizontal lines move further apart as the stress, and hence the strain increases. If we keep the stress constant and cut a notch, or crack at the edge of the material, the strain pattern varies as the crack is successively deepened. We can see that, in Figures 9.5c, 9.5d and 9.5e, the crack

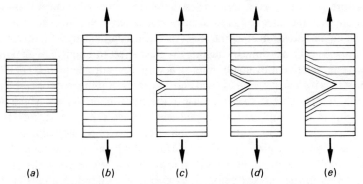

(a)   (b)   (c)   (d)   (e)

Figure 9.5: Strain relaxation occurs behind the tip of the growing crack.

widens and the strained material above and below the crack tends to relax; the lines on the surface close up together behind the crack because, as it widens, it allows some of the strain in the bonds to be relieved. As the depth of the crack increases, the area of the material over which strain relaxation occurs also increases. But, as the strain patterns suggest, the area of strain relaxation and hence the amount of strain energy released actually increases more rapidly than the depth of the crack. When the crack is small (Figure 9.5c) the amount of strain energy released is also quite small, but as the crack length doubles (Figure 9.5d) and then trebles (Figure 9.5e) the area of strain relaxation swept out behind expands increasingly more widely, rather like the ripples spreading out behind a boat—in fact, as we can see, the area has considerably more than trebled in Figure 9.5e.

We are now in a position to be able to examine the crack propagation process in terms of an energy balance. Firstly, energy has to be supplied to create the new surfaces and the amount of energy needed is proportional to the length of the crack. But as the crack grows it releases strain energy at an ever increasing rate. There will come a point, at some critical length, where the strain energy released by a further step in growth will be greater than the surface energy needed for this to occur; sufficient strain energy is then released to pay for the new surfaces and the crack will be able to propagate on its own.

So we can now see that there is a critical length for the crack above which it will tend to grow. The critical length will of course vary depending on the stress level in the material. If the stress is high then a relatively small crack will be able to propagate; conversely, large cracks will propagate at relatively low stresses. The strength of the material is therefore controlled by the size of the largest crack which it contains. Calculations based on these ideas suggest that the size of the cracks which are responsible for the normal strength of glass are of the order of a thousandth of a millimetre deep.

We can put these principles to good use if we want to cut a sheet of glass to size; by making a relatively large crack by scratching the surface with a glass cutter we can quite easily break the glass sheet to the right size.

Glass is not normally subjected to high stresses and accidental surface scratches are therefore generally not a serious restriction to its use. But sheet glass in special applications is sometimes required to have resistance to surface crack propagation and it can be *toughened* against the effect of surface damage by an industrial process called tempering. This is done by heating the glass sheet to a high enough temperature to allow some relative movement in its structure. The surface is then rapidly cooled with air jets; as it cools it becomes rigid and contracts. But the interior is still relatively mobile and its structure is able to adjust to the contraction in the surface. As the interior begins to cool and contract it pulls the surface skin into compression which cannot be relieved because the surface material is no longer hot enough to readjust its structure. When the glass is cold the surface skin is then permanently locked into compression. If we apply a tensile load it is necessary to overcome the built-in compressive stress in the surface before it even begins to develop a tensile stress; the applied load needed to cause the propagation of any surface cracks is therefore increased and, in effect, the glass is strengthened.

So far we have only thought about surface defects in glasses. But similar arguments apply to crystals too. Firstly, it is of course possible to damage the surface of a crystal in a similar way to glass. But crystals also have surface defects because of the way in which they grow; we have already seen that they are built up from layers of atoms, or ions or molecules deposited on their surface. A microscope will often show steps on the surface of a crystal which correspond to the edges of uncompleted layers. These steps cause stress concentrations in a similar way to surface cracks. If all the surface defects are removed, for example by polishing or even by partly dissolving the surface, then the strength of the crystal may be increased.

In a few rather special cases, ceramic crystals with carefully prepared surfaces have been persuaded to show ductility. Figure 9.6 is a simple two-dimensional model which shows that there can be adjacent planes in an ionic structure which do not lead to the electrostatic repulsions shown in Figure 9.2; if the crystal surface is free from stress concentrations then, in a few crystal structures, slip can sometimes occur across planes of this type.

The effect of stress concentrating cracks applies to all materials. So why is it that we cannot generally scratch the surface of a metal sheet with a glass cutter and then break it like glass? The simple answer is that, although stress concentration does occur at the tip of a sharp crack in a metal, plastic deformation takes place locally around it and, in effect, the crack tip can be regarded as being blunted in the process; this has the effect of reducing the stress concentration and the mechanism for further crack growth is inhibited. Cracks certainly do grow in metals but there is evidence to suggest that, even in brittle metals, a considerable amount of energy is consumed in plastic deformation around the tip of a growing crack.

So now we have a picture of how the high resistance to slip or plastic deformation means that the strength of ceramic materials in tension is sensitive to the presence of

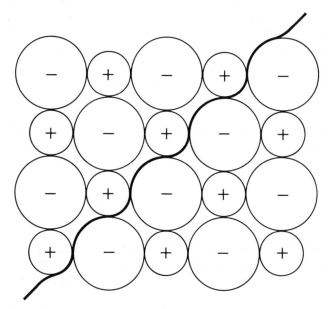

Figure 9.6: A possible slip direction in a two-dimensional model of an ionic structure.

cracks. But ceramics are generally relatively strong in compression; this is also partly because of their resistance to plastic deformation. If a ductile material is subjected to a compressive stress then it will tend to be squeezed out sideways like putty. If a ceramic material is compressed then the nature of its chemical bonding forces tend to restrain it from responding in the same way because the scope for slip is reduced. But furthermore, cracks under compression tend to behave differently from cracks under tension; for example, if the tensile force in Figure 9.5c is replaced by a compressive force then the crack will close and the load can then be transferred across it—under these conditions there will be little incentive for the crack to grow and the material will be able to support a relatively high compressive load.

Ceramics in everyday use are very often composite materials; that is to say they contain several different components which are bound together to form a solid mass. These composites are therefore more complex than the relatively simple crystals and glasses that we have so far been considering. These are two particularly important groups of materials of this type that we should think about.

The first is based on clay, and includes bricks, drainage pipes, tiles and also china and porcelain. The clay is sometimes mixed with other minerals and then it is shaped and finally fired at high temperature. The precise choice of raw materials used and their relative proportions are varied depending on the finished product required.

The firing process is generally carried out somewhere in the region of 800–1450°C but the details of the various processes and the chemical reactions which occur during firing are beyond the scope of this book. Nevertheless we should consider, in broad terms, the internal structure of this type of material. Firing converts the raw materials into a hard composite ceramic which contains both crystalline and glassy components. In an idealised model of the structure, crystalline particles are dispersed in a matrix of glassy material; but there is often a significant proportion of voids or pores as well. By varying the firing conditions it is possible to alter the size of the crystalline regions, the proportion of glass formed and also the porosity.

Because of their complex structures and because of the large number of manufacturing variables, the strength of this type of ceramic is not as well understood as for the simpler materials described earlier. However, a decrease in tensile strength is certainly seen as the porosity increases. The pores reduce the effective load-bearing cross-sectional area and we would expect a decrease in strength for this reason; but it also seems probable that they act as stress concentrators rather like the surface cracks in glass.

The second type of composite ceramic material that we need to consider is concrete. This is perhaps one of the most widely used and important of all engineering materials. It is also one of the most complex, and there are still aspects of its internal structure and consequent behaviour which are not fully understood. Again, this is not made any easier by the wide range of raw materials and techniques that are involved in its use.

A simplified picture of the internal structure of concrete is given in Figure 9.7. Firstly it contains coarse aggregate particles, which are often gravels having a particle size of up to several centimetres. There are also smaller particles of fine aggregates (often of sand) which, together with hardened cement, form a mortar which fills in most of the space between the coarse aggregate particles. The whole structure is bound together by the hardened cement; the cement itself has a very complex microstructure which is still far from perfectly understood.

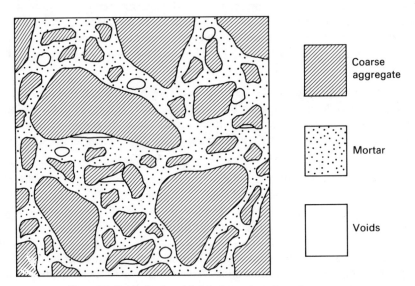

Figure 9.7: An idealised model of the internal structure of concrete.

In addition to these basic components, there are also air voids trapped in the concrete during the initial mixing and pouring operation. Furthermore, water sometimes tends to separate at the top surface of a freshly mixed cement paste; in fresh concrete this water can become trapped underneath the larger aggregate particles and, when the concrete has hardened, voids remain. Shrinkage of the hardened cement paste can also lead to the formation of cracks at the interface between the cement and the aggregate particles.

So we can find a variety of cracks and flaws inside concrete and its seems probable that these defects act as stress concentrators which reduce its tensile strength.

The simplest way to get round this problem of low tensile strength is to keep concrete in compression since, as we might perhaps expect, it is much stronger under a compressive load. But this is not always possible. For instance, in a beam loaded as in Figure 9.8 there will be a tensile stress in the convex face (at the bottom of the beam in

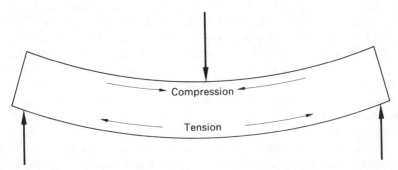

Figure 9.8: A tensile stress is developed in the lower part of the centrally loaded beam supported at both ends.

the figure) where the loading force is tending to make it stretch. If the load is increased then fracture will begin at this face when the tensile stress becomes great enough. The concave face will be in compression but, since concrete has a relatively high compressive strength this will not be liable to fail.

Failure of the beam at the face under tension can be prevented by setting steel reinforcing rods inside the lower part of the beam so that they carry the tensile stress. Sometimes the reinforcing rods are pre-stressed in tension before the concrete is poured around them. When the concrete is hard the external tension is released and the rods pull the concrete surrounding them into a state of compression. The effect of this is analogous to the compressive stress in the surface of tempered glass; the built-in compression must be overcome before a tensile stress can be developed.

To summarise, this chapter has shown how the nature of the chemical bonds in ceramic materials inhibits the slip processes which can occur relatively easily in metals. Ceramics therefore tend to show brittle behaviour and they are generally susceptible to the presence of defects which act as stress concentrators. Since many useful ceramic materials inevitably contain defects of this kind their tensile strengths are often relatively low.

10.

The Nature of Rubbers and Plastics

We have already met the idea that rubber and plastics are based on covalently bonded carbon atoms but we have now reached the point where we must explore this idea in more detail.

In Chapter 6 we saw how the non-directional and non-specific nature of the metallic bond enabled us to think about metal crystal structures in terms of packing uniform spheres together. We saw that we could treat ionic crystal structures in a similar way but we took into account the effect of the oppositely charged ions having different sizes; furthermore we had to remember that similarly charged ions would not come into contact with each other. We saw that covalent crystal structures also exist but that, in this case, the type of bonding imposes directional constraints; in the diamond structure for instance, each carbon atom is surrounded tetrahedrally by four neighbours and the covalent bond between any two atoms is quite specific.

But the directional and specific nature of the covalent bond has a particularly important implication; it enables molecules to exist independently. For instance, a methane molecule consists of a single carbon atom covalently bonded to four hydrogen atoms by means of four orbitals directed towards the corners of a regular tetrahedron in which the carbon atom is centrally located (see Figure 10.1); this molecule now has no need to form further primary bonds and is therefore capable of independent existence. An ethane molecule consists of two carbon atoms joined by a covalent bond; the remaining three orbitals on each carbon atom form covalent bonds with hydrogen atoms and, again, the molecule is a stable entity on its own.

In contrast to this, the collective and non-specific nature of the metallic bond does not lead to discrete molecules; a metal crystal contains an indeterminate number of metal atoms. The total number of ions in an ionic crystal is also indeterminate, although the relative number of different types of ion must be correct to make the crystal electrically neutral.

Of course we must not forget that, in crystals such as diamond, the covalent bond leads to structures of indeterminate size too; but the point that particularly concerns us in this chapter is that it can also lead to the formation of discrete molecules. We have already seen that secondary bonding forces must be involved for a collection of discrete molecules to exist in the liquid or solid state; for instance, water and ice rely on hydrogen bonding and both liquid and solid methane rely on van der Waals forces. In this chapter we shall see that rubber and plastics consist of chain-like covalent molecules with secondary bonds providing attractive forces between them.

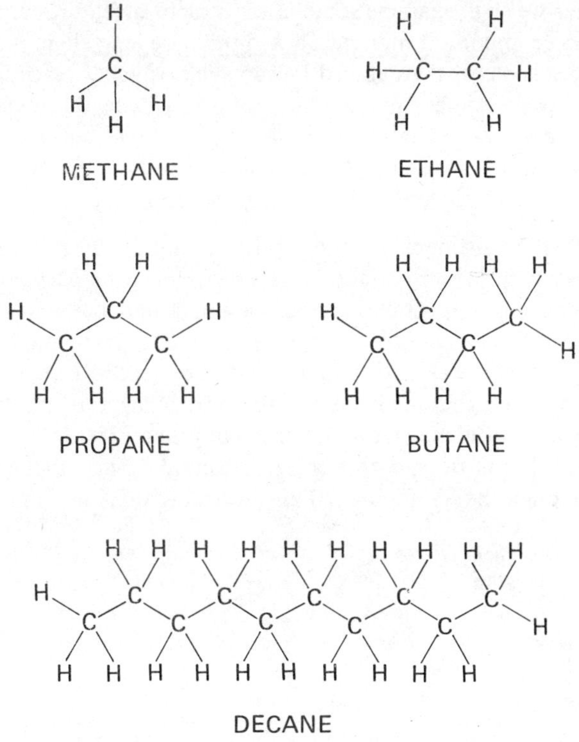

METHANE

ETHANE

PROPANE

BUTANE

DECANE

Figure 10.1: Some simple hydrocarbon molecules

Figure 10.1 shows that methane and ethane are the first two members of a series of chain-like hydrocarbon molecules; by successively inserting —CH_2— groups into a chain we can make it as long as we like. But the long chain molecules that we shall be discussing in this chapter are generally more complex and very much longer than the hydrocarbon chains in Figure 10.1. In fact they are often hundreds or even thousands of carbon atoms long; and sometimes there are side chains branching out of the main backbone chain like branches out of a tree. Very often there are atoms, or groups of atoms, other than hydrogen attached to the backbone carbon atoms; in P.V.C. for example, some of the hydrogen atoms have been replaced by chlorine atoms. Sometimes there are atoms other than carbon in the backbone itself; for instance, nitrogen atoms are built into the backbone chain of nylon. The secondary bonding forces which arise between neighbouring chains have an important effect on the properties of these materials; there will always be van der Waals forces tending to hold the chains together but, by building permanent dipoles onto the chains, the attractive forces between them will be increased. Furthermore, covalent cross-links can be used to join adjacent chains together permanently. All this implies that there is considerable scope for molecular engineering and, in fact, modifications like these are used to adapt the properties of materials of this kind for particular uses.

In this chapter we shall examine some of the structural features and properties of several of the more important materials based on these long chain molecules. Later we shall consider some of the detailed ideas which explain, for example, how the properties of one plastic differ from another; but before doing this we should think, in general terms, about how the nature of these long chain molecules can lead to rather special behaviour which particularly distinguishes these materials from the metals and ceramics that we considered in the previous two chapters.

One property that distinguishes most plastics and rubbers from metallic and ceramic materials is their lower stiffness. As an extreme example, some rubbers can be stretched very easily to several times their original length but will return to it when released; some plastics, when hot, will do the same. This is quite different from the behaviour we expect of metals and ceramics since they generally break at much lower extensions and need a higher applied stress to do so; it suggests that the elasticity of rubber relies on some mechanism that we have not yet considered.

The key to this lies in the fact that rotation can take place about the axis of the carbon–carbon single bond. Figure 10.2*a* illustrates that one end of the ethane

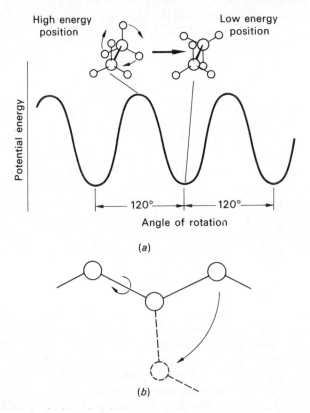

Figure 10.2: Bond Rotation: (*a*) shows the variation in potential energy of the ethane molecule as one end is rotated about the carbon-carbon bond; (*b*) illustrates that rotation of this kind in a polymer chain leads to *bending* of the chain at that point.

molecule can rotate relative to the other rather like a propellor. But there are, in fact, barriers which tend to oppose this rotation. These are due to slight repulsions which arise when the three hydrogen atoms at one end are exactly aligned with the three at the other; this, of course, represents a position of relatively high potential energy. Conversely, there are low energy troughs which occur when the hydrogen atoms at either end are out of alignment. The figure shows that, in rotating one end of the molecule through a full circle, there are three energy barriers separated by low energy troughs. In the case of ethane, the energy needed to overcome these barriers is very low—much lower than the energy needed to break the two carbon atoms apart. We would therefore expect to be able to rotate one end of the molecule relative to the other (about the carbon–carbon bond axis) much more easily than to break the bond apart; in fact, at normal temperatures the molecule has sufficient thermal energy to overcome these rotational barriers by itself, without mechanical assistance from outside.

We can extend this idea to explain the elasticity of rubber; to do this we need to remember that rubber consists of carbon chains which are much longer than those in Figure 10.1 but basically rather similar. We should also note that the chains in the figure are shown as idealised conformations extended in a straight line, although within the chains the carbon atoms have a zigzag arrangement because of their tetrahedral bond angles; but, because it is so easy to rotate each individual carbon-carbon bond about its axis and thereby *bend* the chain at this point (see Figure 10.2*b*), we would expect to find the long carbon chains in rubber as randomly twisted conformations all entangled amongst each other.

It is perhaps easiest to begin thinking about the behaviour of rubber by imagining that we have isolated a single chain from all the rest; this chain might, for example, look something like that shown in Figure 10.3. Because of the small energy barriers, we would expect continuous rotation about the bonds—along the entire length of the chain—under the influence of thermal energy at normal temperatures. We would

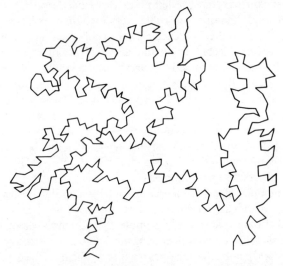

Figure 10.3: A schematic view of one possible instantaneous conformation of a single rubber chain.

therefore expect the chain to be constantly wriggling about and changing its shape; and the higher the temperature, the more vigorous we would expect this thermal wriggling to be. For our large isolated chain there is a very large number of possible shapes or conformations that it could adopt at any one moment because there are so many combinations of low energy positions for all the carbon-carbon bonds in the chain. On statistical grounds it can be shown that most of these possible conformations tend to be fairly compact, as in Figure 10.3; but there are some which are more elongated and, in the extreme case, there is just one in which the chain is fully extended in a straight line in the same way as the decane molecule shown in Figure 10.1. So, although the chain will sometimes wriggle into the more elongated conformations, these are relatively rare; it will tend to spend most of its time more or less bunched up because the compact conformations are more numerous.

We can now see how a rubber chain should respond to an applied stress. If we pull on each end we would expect it to untwist and straighten by bond rotation along its length; if we then release the ends we would expect that thermal motion would soon make it tend to wriggle back to a more compact and statistically more probable form.

But we should of course remember that the single rubber chain that we have been considering is actually surrounded by many neighbouring chains which tend to restrict its movements; in fact, the overall deformation of the rubber will depend on co-operative rotational movements in relatively small adjacent segments of neighbouring chains throughout the material. Nevertheless, the basic argument is still valid because we can still think of the segments as adjusting themselves to an applied stress and then returning to more compact conformations when the stress is released.

The elasticity of rubber is therefore explained by a totally different mechanism from the elasticity of the more rigid materials that we have considered in earlier chapters; now we can see, in principle, how rubber can be stretched to very large extensions by a relatively small stress and how it returns to its original length when the stress is released.

But many plastics in everyday use, like polystyrene and Perspex, normally tend to be harder and more rigid than ordinary rubber although they are based on similar long chain molecules. Where is the difference between rubber and these harder plastics?

We already have one clue, but we can find another by thinking about what happens to a piece of rubber when it is cooled to a very low temperature. When it is extremely cold its properties are quite different; it becomes hard and brittle, and if it is hit with a hammer it may shatter into pieces. Temperature therefore has an important effect. High temperatures lead to increased thermal agitation of the chain segments and low temperatures lead to brittleness. By cooling the rubber, what we have done is to reduce the thermal energy until bond rotation can no longer occur; the chain segments are then frozen into fixed conformations which cannot easily be untwisted by an applied stress and so the rubber becomes hard and rigid.

There are a number of experimental methods which demonstrate the effect of freezing out the rotational movements of chain segments. One of these involves measuring the thermal expansion of a sample of the material as the temperature is raised, and other methods involve measuring mechanical properties such as the elastic modulus. The general picture that emerges from experiments of this kind is illustrated in Figure 10.4. The first graph shows the thermal expansion effect; we can see that the volume initially increases quite slowly with temperature, but there is a kink in the graph and as the temperature is raised above this then the volume increases more rapidly. In fact, this kink corresponds to the temperature below which bond rotation is frozen; below it, thermal motion is confined to vibration of atoms about equilibrium positions in the solid material—the slight increase in volume with temperature in this region is simply due to the increasing vigour of the vibrations as the temperature is increased. When the temperature corresponding to the kink is reached, then the thermal energy is sufficiently high for bond rotation to begin and for the chain segments to start wriggling about. But, rather like a skipping rope, the wholesale movement of a chain segment requires plenty of room and, as the temperature is raised further, the extra volume swept out by the increasingly vigorous movements of the segments will increase relatively rapidly.

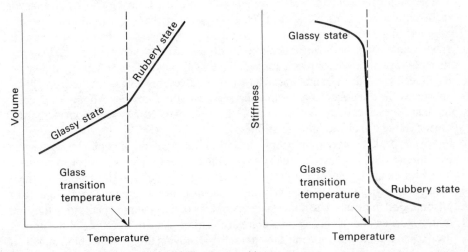

Figure 10.4: The glass transition temperature.

The second graph in Figure 10.4 shows that the stiffness, or elastic modulus of the material shows a sharp decrease above a particular temperature; not surprisingly perhaps, this also corresponds to the temperature at which free movement of the chain segments begins. Above this temperature rubberlike behaviour is observed, but below it the restriction of bond rotation makes the material relatively rigid and hard. So, at lower temperatures its properties will tend to be more like those of glass than of rubber; Perspex and polystyrene, for instance, exist in this glassy state at room temperature and are therefore normally regarded as hard and brittle materials.

This temperature, at which the transition from glassy to rubberlike behaviour occurs, is called the glass transition temperature.

We can now see that the properties of these long chain materials are very much dependent on whether they are measured above or below the glass transition temperature; it therefore follows that the usefulness of a particular material for a specific application will generally depend on whether its glass transition temperature is above or below the normal temperature at which it is to be used.

Before we go on to think about how the structural features of the chains affect the glass transition temperature, there is a complication that we should briefly consider; namely that the properties of these materials are dependent on time as well as on temperature. A good illustration is given by bouncing putty, which is sold in toyshops. Bouncing putty can be moulded into shape just like ordinary putty; but if it is rolled into a ball this will bounce just as if it were made of rubber—and yet, if it is left to stand, it will slowly flow into a puddle under its own weight like a very viscous liquid. The reason for this strange behaviour is that bouncing putty contains long chain molecules which, under slow loading conditions, have enough time to disentangle themselves and to actually slide past one another so that the material deforms plastically; under fast loading conditions, however, the chains do not have sufficient time to flow in this way and the material then behaves elastically.

The chains in ordinary rubber are normally joined to one another by permanent cross-links at intervals along their length to prevent them from slipping past one another. But the mechanical properties still vary depending on the rate at which the rubber is deformed because the chain segments need a certain amount of time to adjust their conformations as a stress is applied. If the rubber is stretched relatively slowly the chain segments have enough time to adjust themselves; but if it is stretched rapidly the shortened time scale does not allow the segments the same opportunity to accommodate themselves to the applied stress—the result of this can be seen as an increase in the measured elastic modulus, or stiffness.

We can begin to see that the behaviour of these materials is quite complex. A thermal expansion experiment carried out by raising the temperature very slowly (over several hours for example) will give a fairly accurate indication of the temperature at which bond rotation begins. But a relatively high speed mechanical test, such as an impact test of some kind, may not detect thermal movements of the chain segments until they are quite vigorous and occurring on a comparable time scale to the effect of the impact itself; even though the segments may actually be moving they may not have sufficient time to adjust themselves to a very rapidly applied stress. A high speed test may therefore indicate that the transition from the glassy state to the rubbery state occurs at a higher temperature than the kink on the volume versus temperature graph would suggest. This means that, at a given temperature, a particular material may normally be regarded as tough and resilient although, in a high speed impact situation, it may in fact fracture in a brittle manner.

Although we should bear all this in mind and remember that the behaviour of these materials is really rather complicated, we can nevertheless usefully discuss, in quite simple terms, some of the more important features of their structures which influence their properties.

Firstly we should note that the length of the chains will be important. In fact, we usually think in terms of the average length because the chains in any one particular material normally cover a range of different lengths. To illustrate the effect of the average chain length we will think about an imaginary experiment in which we begin with a material which contains long chains and look at what happens if we successively reduce their size.

If the material is above its glass transition temperature and has very long chains we would expect it to show rubberlike elasticity; even though it may not contain cross-links between the chains there should be a sufficient number of points of entanglement between them to act, in effect, as temporary cross-links provided that any stress is not applied for too long. (If a stress is maintained for a long period the restoring tension will tend to decrease gradually because the chains will then have sufficient opportunity to slip past one another.) If the average length of the chains is reduced then the material will tend to behave more like a liquid; it may, in fact, go through an intermediate stage where the chains are short enough to allow it to flow like a very viscous liquid but still long enough for some entanglement to cause rather weak rubbery retraction if it is stretched and quickly released. Finally, if the chains are reduced to a very short average length, we would expect to see normal liquid behaviour.

Figure 10.5 shows how we might expect the glass transition temperature to vary as the chain length is reduced in our imaginary experiment. The figure illustrates that, at

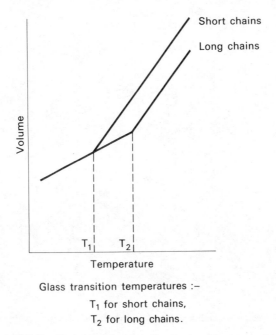

Glass transition temperatures :–

T_1 for short chains,

T_2 for long chains.

Figure 10.5: The effect of chain length on the glass transition temperature.

any one temperature in the rubbery region, a particular weight of short chains will occupy a greater volume than the same weight of long chains. The reason for this is that a terminal segment at the end of a chain is less restricted in its thermal motion than a segment in the middle; a terminal segment is only attached to the rest of the chain at one point whereas a segment in the middle is attached by both its ends and is therefore less free to move—terminal segments therefore tend to wriggle more vigorously than segments in the middle and so they occupy more volume. A given weight of short chains will obviously contain a larger number of free ends than the same weight of long chains; in the rubbery state a material containing short chains will therefore occupy more

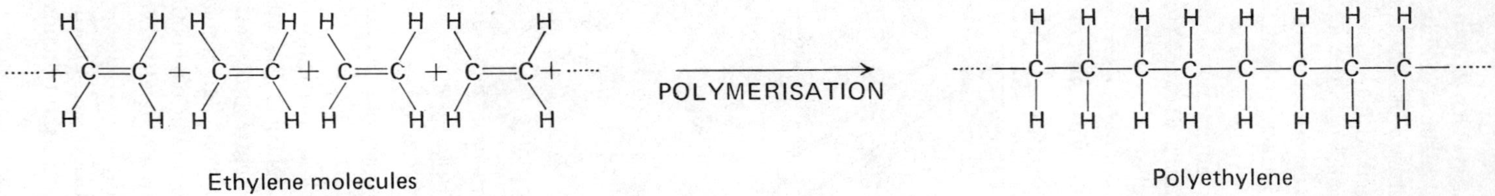

Figure 10.6: The polymerisation of ethylene to form polyethylene.

volume than a long chain material at the same temperature. In the glassy state there is no difference between the volume of long chains and short chains because the movement of segments has been frozen. The result of all this is shown in Figure 10.5; we can see that the rubber line for the short chains intersects the glass line at a lower temperature than the rubber line for the long chains—the glass transition temperature is therefore lower for the short chain material.

So now we have a simple picture of how the properties of these materials can depend on the length of their chains. But so far we have not thought about the effect of the chemical structure of the chains themselves; we have simply considered them as being built up from covalently bonded carbon atoms and have paid no attention to the effect of other atoms attached to them.

Materials like plastics and rubbers are called polymers. The word *polymer* is applied to materials which can be regarded as being made by joining together small molecular units, called *monomers,* in a repeating pattern to form large molecules; the process of forming polymers in this way is called *polymerisation.* To take an example, polyethylene (commonly called polythene) is a polymer which is made from ethylene. Figure 10.6 shows that, in effect, the polymerisation of ethylene molecules involves the opening of their double bond so that they can join end-to-end in large numbers to form long chains.

It is beyond the scope of this book to discuss polymerisation in any detail but it should be quite clear that, by using different monomers, we can vary the chemical structure of the completed polymer chain.

Figure 10.7 gives a series of important polymers in everyday use. The series begins with polythene where, as we already know, the repeating unit derived from ethylene is $—CH_2—CH_2—$. We can think of the structure of polypropylene in terms of the polythene chain; in effect, one of the hydrogen atoms of the polythene repeating unit has been replaced by a $—CH_3$ group of atoms—furthermore, in P.V.C. the hydrogen atom has been replaced by a chlorine atom and in polystyrene by a group of atoms called a benzene ring. Finally, in the case of Perspex, two of the hydrogen atoms of the polythene repeating unit have been replaced by groups of atoms.

Figure 10.8 gives us an idea of the relative values of the glass transition temperature for the members of this series of polymers; assuming there are no differences due to testing rate effects or chain length, we can see that the glass transition temperature tends to increase as the substituent attached to the backbone chain becomes larger and more complex—and Table 10.1 suggests that this increase in size and complexity also leads to a general increase in strength and stiffness. This is what we should expect; if one of the hydrogen atoms in the ethane molecule in Figure 10.2*a* is replaced by a larger atom, or a group of atoms, then the barriers to rotation will be increased. Furthermore, we should also expect that the presence of larger substituents on the backbone chain would lead to an increase in the mutual restraining effect of neighbouring chains in the polymer.

In short, these factors increase the glass transition temperature and both strengthen and stiffen the polymer by reducing chain mobility. But there are less obvious factors involved too. For instance, at first glance we might be surprised by the

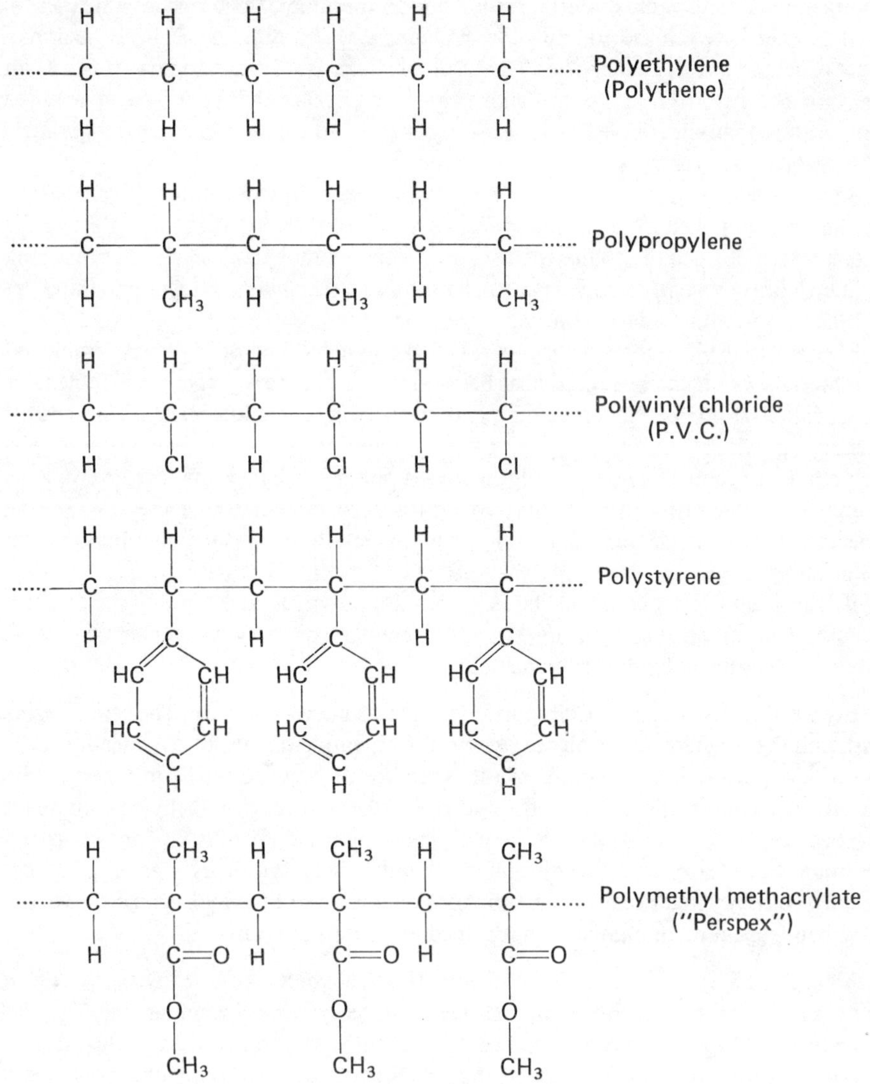

Figure 10.7: Some important thermoplastic polymers.

apparently rather high glass transition temperature of P.V.C. and its rigidity and strength; both Figure 10.8 and Table 10.1 suggest that the effect of the single chlorine atom is comparable with that of the larger and more complex benzene ring in polystyrene. To understand this we should remember, from what we saw in Chapter 4, that the carbon-chlorine bond is permanently polarised. This leads to relatively strong attractive electrostatic forces between adjacent chains; these forces are then responsible for the high glass transition temperature and they make P.V.C. a harder and more rigid material than we would expect from van der Waals forces alone.

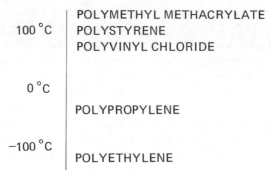

Figure 10.8: A simple comparison of the glass transition temperatures of some important thermoplastic polymers.

Material	Tensile Strength	Stiffness	Elongation at break
Thermoplastics:			
Polyethylene —			
a) Low crystallinity	Low	Low	High
b) High crystallinity	Medium	Low—Medium	High
Polypropylene	Medium	Low—Medium	High
Polyvinyl chloride —			
a) Plasticised	Low—Medium	Low—Medium	High
b) Unplasticised	High	Medium	Low—Medium
Polystyrene	High	Medium	Low
Polymethyl methacrylate	High	Medium	Low
Thermosetting resin:			
Phenolic	High	High	Low

Table 10.1: A simple comparison of the properties of several important plastics.

But many commercial applications, such as clothing for example, require P.V.C. to be made soft and flexible. This can be done by the use of plasticisers which are added to the polymer (see Table 10.1). These plasticisers are often relatively small molecules which themselves contain permanently polarised bonds in their structure; when plasticiser molecules are blended with P.V.C. they are attracted to the polarised parts of the polymer chain which are then, in effect, *neutralised* with the result that the attractive forces between adjacent chains are reduced. Furthermore, the penetration of the plasticiser molecules between the chains will tend to push them apart; this, in itself, will of course tend to reduce the attractive force between them.

So far we have considered a polymer to be an amorphous material consisting of a mass of chains in random conformations all haphazardly entangled amongst each other. But this picture is often far from true because some polymers, including polythene and polypropylene, can usually *crystallise* quite easily. Figure 10.9 illustrates, in a highly idealised way, how sec dary bonding forces are able to pull adjacent chains into stable, low energy arrangements in which they are closely packed parallel to one another. Within these regions crystalline order exists which is similar to that already discussed in earlier chapters; in this case, however, the relatively cumbersome polymer chains are not generally able to arrange themselves along their entire length to give a completely crystalline material. Single polymer crystals

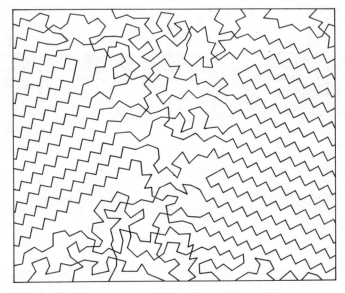

Figure 10.9: A schematic model of a polymer structure showing crystalline and amorphous regions.

can sometimes be prepared by very special laboratory techniques; but we can think of the structure of ordinary crystalline polymers in terms of small crystalline regions which are dispersed in a matrix of amorphous material. As Figure 10.9 suggests, a particular chain may, in fact, pass through several regions of crystallinity; and there is evidence to suggest that a single chain may fold back on itself a number of times before it leaves one of these regions.

Crystallinity in polymers affects their properties; firstly, we might expect the crystalline material to be stronger and stiffer than the amorphous material but, in addition to this, the crystalline regions tend to restrict the mobility of the polymer structure as a whole. A highly crystalline polymer will therefore tend to be stronger and more rigid than a similar material which contains a higher proportion of amorphous material.

Polythene has a simple chain structure and is normally able to crystallise very easily; this is important in determining its useful properties because its glass transition temperature is well below room temperature. Although we might expect the amorphous regions in polythene to be normally in the rubbery state, the crystalline regions do not *melt* until they are well above room temperature and they therefore give the polymer its rigidity at normal temperatures. The extent to which crystallisation occurs in polythene is affected by the *branching* of side chains out of the main backbone chain. The chain structure of the side branches is essentially just the same as the main backbone but their presence does interfere with crystallisation. A highly branched polymer will therefore tend to be less crystalline and so would be expected to be weaker and more flexible than a less branched and more highly crystalline material (see Table 10.1).

So now we can see how secondary bonding forces and the presence of crystalline regions both increase the rigidity and strength of polymers by tying together adjacent chains. Finally, we shall think about the effect of forming permanent covalent cross-links between the chains in a polymer structure.

Earlier in this chapter we noted that we can prevent rubber chains from slipping past one another, under an applied stress, by forming permanent cross-links between them at intervals along their length. If we increase the number of cross-links between the chains then the rubber becomes harder. Ebonite, which is a highly cross-linked rubber, is quite hard and rigid because chain mobility is very much restricted by the larger number of cross-links.

There is a wide range of useful materials, called thermosetting resins, which have cross-linked structures. Well-known examples of these include polyester, phenolic and epoxy resins; polyester resins are particularly well-known because of their wide use in making fibre-glass mouldings such as boats and motor vehicle bodies. A detailed description of the chemical structure of thermosetting resins is beyond the scope of this book; nevertheless we must understand that they are essentially different from the *thermoplastic* polymers which we discussed earlier. Thermoplastics do not contain cross-links and therefore become *plastic,* and easily moulded, at high temperatures; the effects of secondary bonding and the presence of crystalline regions can be overcome by heat but they are then re-established by cooling the material down again. Polythene bowls, for example, are made by injecting the hot, molten polymer into a mould and allowing it to cool.

But, as their name implies, thermosetting resins are set or hardened by the application of heat; the chemical reactions which lead to cross-linking may require fairly high temperatures although, for some resins, room temperature is enough. Once the permanent cross-links have been formed the resins cannot be remoulded; this means that the chemical cross-linking reactions must be carried out during the moulding operation. For instance, in making a polyester resin/glass fibre moulding, the chemical components of the resin are mixed just before it is used to impregnate the glass fibre, which is generally in the form of a cloth. After soaking into the glass fibre cloth the resin then hardens, normally at room temperature, and the resulting material is a composite of glass fibres dispersed in a continuous matrix of cross-linked resin.

Hardened thermosetting resins tend to maintain their rigidity at rather higher temperatures than thermoplastics because their structures are comparatively rigidly held by the cross-links; furthermore the cross-linking means that, although these materials are often relatively strong and rigid compared with other plastics, they tend to be brittle.

We can now summarise, in very broad terms, some general points which help us to compare polymers with metals and ceramics.

Firstly, their chain structure is based on covalent bonding (which means that they are often useful electrical insulators). But the relative movement of the chains in a polymer structure can generally occur without much difficulty and polymers therefore tend to be weaker and less stiff than metals and ceramics; for instance, the overall range of tensile strengths for the materials in Table 10.1 lies roughly between $10\text{--}70 \times 10^6$ N/m^2 and the elastic moduli between about $100\text{--}7000 \times 10^6$ N/m^2—this compares with corresponding values of about $35\text{--}170 \times 10^6$ N/m^2 and $70\,000 \times 10^6$ N/m^2 for ordinary glass. Furthermore, the response of polymers to an applied stress tends to be very highly dependent on temperature and on time; where metals and ceramics simply rely on the stretching or compression of chemical bonds, their elastic response is more or less instantaneous.

The strength and rigidity of polymers can be increased to some extent by structural modifications which tend to immobilise the chains; but brittleness is sometimes the penalty which has to be paid for this—by reducing chain mobility, the scope for plastic deformation is often diminished.

11.

Conclusion

In Chapter 1 we began by discussing three criteria which help us to describe the behaviour of any material; and we identified these as stiffness, strength and toughness. Now we have reached the point where we have a broad understanding of how the behaviour of different materials is influenced by the nature of the chemical bonds which operate within them. For instance, we have seen that stiffness and strength both stem from the cohesion due to chemical bonding.

Provided that any applied mechanical force is not too great, then a solid material deforms elastically and will return to its original shape and size when the force is removed; and we have seen that the elastic modulus, or stiffness of a material is a measure of its reluctance to be stretched. Whether this involves the distortion of a crystal lattice or the wriggling motion of polymer chains, stiffness is dependent upon the chemical nature of the material.

But the strength of a material also depends on its chemical nature and, in theory, it should be possible to calculate how much stress can be applied to it before it breaks. However, the presence of imperfections, such as dislocations and cracks, very often weakens the internal structure so that the material undergoes plastic deformation or brittle fracture long before the maximum theoretical strength can be reached.

High strength generally implies a low degree of mobility in the internal structure. The relatively low strength of a pure single crystal of copper is due to the ease with which dislocations can run through its structure. In a similar way, the low strength of polythene can be related to the ease with which the polymer chains are able to slip past one another. On the other hand, there is much less scope for internal mobility in diamond or glass and these materials are intrinsically strong—but they are brittle.

It is unfortunate that, on this basis, high strength generally tends to confer brittleness on a material; conversely, toughness usually implies some inherent weakness. We can follow the argument behind this generalisation by referring to Figure 11.1. A material which possesses a high degree of mobility in its internal structure can readily adjust itself to the effect of a stress concentrator by plastic deformation; in effect the stress concentration is relieved by the blunting of the crack tip due to localised slip processes. In a material which is intrinsically strong, with a low degree of internal mobility, plastic deformation is much more difficult; as we saw in Chapter 9, the stress will tend to be concentrated in the bonds at the crack tip—the crack will then propagate rather than become blunted.

But apart from mechanical properties, the selection of a material for a particular

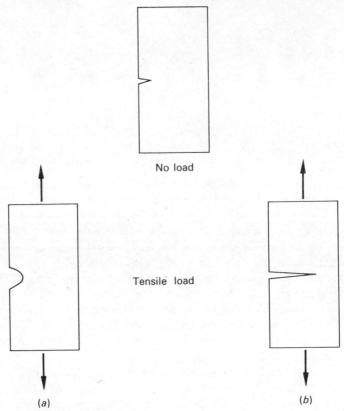

Figure 11.1: (*a*) Plastic deformation enables stress concentrations to be relieved; (*b*) but if this cannot happen, then brittle fracture will ultimately occur.

application generally involves considering other factors, such as cost and the ease with which it can be converted into finished products. It is beyond the scope of our discussion to consider factors of this kind but we should bear in mind that they are usually extremely important and that they can sometimes emphasise the mechanical disadvantages of a particular material. For instance, it may be cheaper to manufacture some product by moulding it in one piece from a plastic rather than by shaping and machining metal components which then have to be assembled—but this presupposes that there is a polymer which has the necessary strength and stiffness for the product to be able to do its job properly in service. To take another example, the development of the aerospace industry has emphasised the need for materials which are stiff and strong but, at the same time, light in weight; a number of ceramics combine lightness with high stiffness and are potentially extremely strong—but their brittleness is a major drawback.

So, for a variety of reasons, materials scientists are paying considerable attention to widening the range of applications of particular materials by trying to find ways of overcoming their inherent disadvantages. And one way that they are doing this is by combining different materials together to form composites which possess special properties of their own that cannot be attributed to their individual components. A good illustration of this principle is given by fibreglass.

A special property of fibreglass is its toughness. Of course, the individual fibres in a fibreglass composite are still brittle; furthermore, the cross-linked resins used to bond them together are often rather brittle themselves, and generally not very strong or stiff either. But the finished composite has sufficient toughness to be used in the form of boat hulls or car bodies. How is this toughness achieved?

To form a simple picture of how fibreglass works, we can regard it as a collection of glass fibres glued together with resin, as shown schematically in Figure 11.2; apart

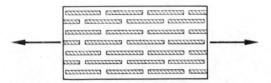

Figure 11.2: In fibreglass, the resin bonds the glass fibres together so that they can act collectively in supporting an applied load.

from acting as a glue, the resin also protects the fibres from surface damage which would weaken them. Since the fibres are bonded together, they can co-operate with one another and act collectively in resisting any force applied to the composite—and, as this implies, the resin is involved in transmitting stress from one fibre to another. When the load is applied to the composite, the strain in the resin as a whole (i.e. the overall extent to which it is stretched) is about the same, on average, as the strain in the fibres. But, because the glass fibres are much stiffer than the resin, they carry most of the load—this idea becomes clearer if we remember that we need to apply a higher stress to stretch glass (with its relatively high modulus) than resin by the same amount.

So now we can see how the fibreglass composite carries a load, but we still have no explanation for its toughness. To understand this, let us imagine that there is a small crack in the continuous resin matrix which is being propagated by the applied force.

Provided that the adhesion between the resin and the glass is not too great then the fibres can act as "crack stoppers". Figure 11.3 shows that, as the crack approaches a fibre, the adhesion at the interface breaks down and the resin tends to split away from the glass—in effect, the crack is blunted and further propagation is then inhibited. Even if the crack is able to find its way past the fibre, as in Figure 11.4, its progress is still made difficult because the fibre forms a bridge which tends to hold the two opposite crack faces together; for the crack to grow any further the faces must be moved apart—and for this to happen, the fibre must be pulled out of the resin against the resistance due to adhesion at the interface. So although neither the glass nor the resin have much resistance to crack propagation on their own, the fibres throughout the composite provide a large number of barriers which resist the progress of cracks—and therefore fibreglass composites tend to be tough.

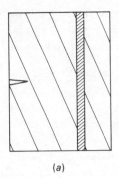 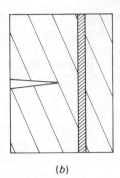 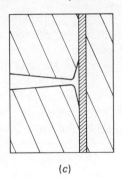

(a)   (b)   (c)

Figure 11.3: Crack propagation can be halted by the breakdown of the resin/glass interface at the surface of a fibre.

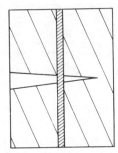

Figure 11.4: Further propagation is restricted by the fibre tending to hold the opposite faces of the crack together.

In the illustration shown in Figure 11.2, the fibres are arranged parallel to one another and the stiffness and strength of the composite is concentrated in this direction; in practice, the alignment of fibres is generally varied to give the composite more uniform properties in different directions.

But fibreglass is just one of a large variety of composites. There are many other instances where the combination of different materials results in improved properties. Again, some of these involve the use of polymers—both thermosetting and thermoplastic—and a wide range of different types of particles and fibres have been used to modify their properties. For instance, the stiffness and wear resistance of rubber may be improved by incorporating carbon black particles into it; on the other hand, polystyrene can be toughened by a dispersion of rubber particles. Another useful combination is obtained by bonding paper fibres with thermosetting resins, and materials of this type form the basis of some household decorative laminates. To take an example of another kind, there are a number of ceramic powders which can be glued together with metallic binders to form extremely hard composites that are used for making cutting tools of various kinds; at the other end of the scale, certain metals can be strengthened by dispersing tiny ceramic particles throughout their structure to restrict the movement of dislocations.

These are just a few from the many examples that could be taken to illustrate the range and variety of different combinations of materials that can be made. But Nature has produced some interesting composites too.

Wood is a highly complex material which contains large numbers of hollow tubular cells packed parallel to one another to give an arrangement rather similar to a bundle of drinking straws. Each individual cell has a complex structure based on an intricate framework made from cellulose, which is a natural polymer. The detailed structure of different woods varies enormously, but the parallel arrangement of the cells can lead to considerable tensile strength in the direction along which they are aligned; furthermore, the fibrous nature of wood provides inherent toughness.

Compact bone is another natural composite material. The load bearing properties of bone depend on the combination of hydroxyapatite, which is a ceramic-like material, and a protein called collagen. The hydroxyapatite takes the form of tiny elongated crystals which are supported, in a matrix largely consisting of collagen, in an analogous way to glass fibres supported in a resin matrix; and in fact, simple calculations suggest that the relatively stiff hydroxyapatite crystals are probably responsible for supporting most of the load in rather a similar way to the fibres in fibreglass composites.

Previous chapters have shown us that the mechanical behaviour of a particular type of material is fundamentally related to its chemical nature; and this applies to its limitations as well as to its advantages. And we have also seen that the behaviour of materials can be altered, to some extent at least, by modifying their internal structure—to take just one example, we have seen that the properties of plastics depend on the detailed structure of the polymer chains. Improvements in our basic materials, due to modifications of this kind, will undoubtedly continue in the future—but it seems likely that much exciting progress will also be made in the development of composite materials.

In developing fibrous composites like fibreglass, materials scientists seem to have been imitating Nature. But there still remains a great deal to be learned about the structure and properties of natural materials. Perhaps future research in this area will provide new ideas for man-made composites?

Revision Questions

1. Atoms may be regarded as being constructed from three types of fundamental particle.
   a) Describe each of these particles.
   b) Explain, in general terms, how the presence of each type of particle can be demonstrated in matter.
   c) Discuss the arrangement of these particles in atoms with brief reference to early experiments on atomic structure.

2. Write brief notes explaining the significance of each of the following in describing the structure and properties of atoms:
   a) the electron;
   b) the nucleus;
   c) atomic number;
   d) ionisation energy;
   e) the electronic configuration of the inert gases.

3. Write brief notes on four of the following, using sketches where appropriate:
   a) the behaviour of electron beams as demonstrated in the cathode ray tube;
   b) isotopes;
   c) the 1s, 2s and the three 2p orbitals of the hydrogen atoms;
   d) the electronic structure of the water molecule;
   e) metallic bonding;
   f) the variation in ionisation energy of the elements with position in the periodic table.

4. By discussing the nature and origin of the component attractive and repulsive forces in each case, show how the net force/distance curve arises for:
   a) the ionic bond;
   b) the covalent bond;
   c) the metallic bond;
   d) the van der Waals bond;
   e) the hydrogen bond.
   Give an example of each type of bond.

5. From their atomic numbers, and by reference to the periodic table (Table 3.4, page 22), identify the elements in the following pairs and deduce:
    a)   the ground state electronic configuration of each element;
    b)   the type of chemical bond which is found between the atoms in each pair;
    c)   the nature of the chemical structure of the resulting material.

	Atomic number	
	Element A	Element B
Pair 1	8	1
Pair 2	8	12
Pair 3	6	17
Pair 4	11	8
Pair 5	19	19

6.   a)   Use sketches to illustrate the nature of the primary bonding in the water molecule and hence show how secondary electrostatic forces arise.
    b)   Explain carefully the nature of the changes which occur as water is converted from the solid to the liquid and finally to the gaseous state.
    c)   Account for the progressively lower temperatures at which magnesium oxide, sodium chloride, water and argon liquefy.

7. Describe the essential differences between gases, liquids and solids in terms of:
    a)   their mechanical properties;
    b)   the balance between the cohesive force due to chemical bonding and the effect of temperature;
    c)   the structural arrangement of their component atoms (or ions or molecules).

8. With due reference to the electronic structures and the chemical bonding forces involved, explain why:
    a)   at room temperature, sodium bromide is a crystalline solid;
    b)   iron is a lustrous, opaque material which conducts electricity;
    c)   the boiling point of water is much higher than that of methane.

9.   a)   With the aid of sketches, show the difference between the body-centred cubic and the face-centred cubic crystal structures.
    b)   Explain the significance of the radius ratio in determining ionic crystal structures.

10. Show, with the aid of a diagram, the general way in which the forces between two chemically bonded atoms vary with interatomic separation and describe carefully how such forces might be expected to be reflected ultimately in the mechanical properties of an ideal material.

Briefly illustrate how this simple model can often fail to explain the mechanical behaviour of real materials in everyday use.

11. Define the types of chemical bonding present in:
    a)　　sodium chloride;
    b)　　diamond;
    c)　　ice;
    d)　　copper;
    e)　　solid argon.

Show how the nature of the bonding in these chemical structures explains the general form of their internal geometry.

12. Carefully explain each of the following phenomena in terms of the scientific principles involved:
    a)　　work must be done to stretch a film of liquid;
    b)　　the elastic modulus of many crystalline solids decreases if they are heated;
    c)　　despite their ionic structure, sodium chloride crystals can be made to show ductility.

13. a)　　The strength of ionic and covalent bonds in ceramics is generally high. Explain why, in spite of this, samples of (i) ordinary sheet glass and (ii) concrete tend to fail under relatively low tensile loads.
    b)　　A technique, common in principle to both sheet glass and to concrete beams, is employed to offset this tensile weakness. Briefly discuss the basis of this technique and show how it can be applied in both instances.

14. Write brief notes on four of the following, using sketches where appropriate:
    a)　　the fundamental basis of Hooke's Law;
    b)　　the sodium chloride crystal structure;
    c)　　slip in the hexagonal close-packed structure;
    d)　　dislocations in metals;
    e)　　the plasticisation of P.V.C.;
    f)　　the internal structure of soda-glass.

15. a)　　Why should a copper crystal normally be so much more ductile than a sodium chloride crystal?
    b)　　'Without hydrogen bonding, water would normally be a gas.' Explain this statement by making due reference to the electronic structure of the water molecule.
    c)　　The structures of both polythene and graphite rely on a combination of covalent and van der Waals bonding, but their properties are greatly different. Discuss the structural differences between the two materials and hence explain how their properties are affected.

16. a)   The elastic behaviour of rubber is fundamentally different from that of metals or ceramics.
(i) Describe the mechanism of rubber elasticity, carefully explaining how it leads to low modulus values and high extensions.
(ii) Explain why rubber becomes brittle if it is cooled below its glass transition temperature.

b)   By referring to the differences in their chain structure, explain why polythene is tougher than polystyrene at room temperature.

17. Carefully outline the main structural and property characteristics that distinguish ceramics from metals and organic polymers.

18. By referring to the internal structures involved, broadly discuss the general type of failure you would expect in each case if tensile tests were performed on the following materials:

a)   an aluminium crystal;
b)   Perspex;
c)   a sodium chloride crystal;
d)   diamond.

19. Describe the structure of polythene and briefly comment upon the significance of crystallinity in determining its properties.

By referring to their basic structural differences, compare and contrast the mechanical behaviour of polythene with that of rubber and thermosetting resins.

20. With special reference to chemical bonding and internal structure, discuss the differences between:

a)   thermoplastic and thermosetting polymers;
b)   the electrical conductivity of metals and plastics;
c)   the toughness of ceramics and metals;
d)   polythene and P.V.C.

21. A number of materials were compared by placing sheets of standard width and thickness on two parallel supports and then allowing a weight to fall from a fixed height onto the centre line between the supports. The materials used were glass, aluminium and polythene. In the test, the glass shattered, the aluminium bent but the polythene remained undamaged.

An observer was heard to conclude that the polythene was stronger than the aluminium which, in turn was stronger than the glass. Carefully explain the nature of his misunderstanding.

22. a)   Explain why the actual strength of glass is generally considerably lower than its theoretical strength.

b)   Explain, in structural terms, how the normal tensile weakness of glass can be offset by (i) tempering and by (ii) making it into fibres.

c)   Glass fibres and cured polyester resin are both brittle materials.
      Glass fibre/polyester resin composites have applications in the manufacture of motor vehicle bodies, boat hulls and in the building industry.
      How are these two apparently incompatible statements reconciled?

23. State whether each of the following is brittle or tough and, in each case, justify your answer by due reference to the internal structure of the material:
   a)   glass;
   b)   annealed copper;
   c)   polystyrene;
   d)   a glass fibre/thermosetting resin composite.

24. Explain the basic reasons why many real materials seldom develop, in practice, their fundamental theoretical tensile strength.

Index